Convair
F-102
Delta Dagger

A Photo Chronicle

Wayne Mutza

Schiffer Military History
Atglen, PA

Book Design by Ian Robertson.

Printed in China.
ISBN: 0-7643-1062-3

We are interested in hearing from authors with book ideas on related topics.

Published by Schiffer Publishing Ltd.
4880 Lower Valley Road
Atglen, PA 19310
Phone: (610) 593-1777
FAX: (610) 593-2002
E-mail: Schifferbk@aol.com.
Visit our web site at: www.schifferbooks.com
Please write for a free catalog.
This book may be purchased from the publisher.
Please include $3.95 postage.
Try your bookstore first.

In Europe, Schiffer books are distributed by:
Bushwood Books
6 Marksbury Road
Kew Gardens
Surrey TW9 4JF
England
Phone: 44 (0)208 392-8585
FAX: 44 (0)208 392-9876
E-mail: Bushwd@aol.com.

Try your bookstore first.

Contents

Acknowledgments

This is as much the story of the people who were involved with the Convair F-102 as it is of the airplane itself. They range from pilots who flew the missions to those who share my fondness for the impressive interceptor. During my research, one of the many people with whom I had the good fortune to correspond was Martin O. "Mo" Detlie, Lt. Col. USAF Ret. Mo kindly availed to me his entire scrapbook, which, to my delight, was a testimonial to his aviation accomplishments. His was a career track that spanned from fledgling aviator, through his days in P-47s, P-80s, the "Starblazers" demonstration team flying F-84s, the F-86, and finally, to the "Deuce," which he not only mastered, but flew in the 1957 Bendix Race.

A similar path was taken by Budd Butcher, one of the first to fly the F-102 operationally, who flew a Deuce through an atomic cloud, led the first F-102 deployment abroad and aspired to the position of Commander, Air Forces Iceland. Others whose contributions were based on their experiences are: Thomas W. Sawyer, Maj. Gen. USAF Ret., who commanded the last active duty F-102 squadron in the Air Force; Leslie J. Prichard, who made the first arrested landing with an F-102; Colonels Kenneth Kurtzman and John M. Patton, commanders of the 31st Fighter Interceptor Squadron; Joseph S. Algranti, who flew a wide variety of NASA aircraft; John M. Fitzpatrick, Convair test pilot; and Paul L. Wagner and Jack Aycock. David Maass provided tremendous insight into the F-102's involvement with the drone program.

Research of any aircraft flown by the U.S. Air Force Air Defense Command inevitably forms a path that leads to Marty J. Isham, David R. McLaren, and Jerry Geer. The knowledge and materials that they so kindly imparted shine like a light throughout this book. Marty provided not only many of the images that appear here, but previewed portions of the manuscript with the critical eye of an ADC expert.

Equally important and most welcome was the generous input of Robert F. Dorr and the endless flow of material from Lennart Lundh, both of whom are accomplished in the field of aviation literature. Also deserving of special mention are William P. Streicher, a Deuce enthusiast who provided many of the rare insignia presented here, and Kirsten Tedesco of the Pima Air and Space Museum, who continually supports my work with material from the vast facility.

Others who gave freely of their time and material to make this account possible, and to whom I am most grateful, are: David P. Anderson, Paul D. Boyer, Jim Burridge, Lt. Col. Anthony J. Christiano of the 160th RW of the New York Air National Guard, Michael Druzolowski, Charles Ellington of the LWEG at Dahlgren, SSgt Stephen N. Fields of the 101st ARW of the Maine Air National Guard, Phillip Friddell, John Guillen, John L. Hairell, David Hansen, Thomas Hansen, Dr. J.D. Hunley of NASA/Edwards, Scott W. Johnson of NORAD/U.S. Space Command, Leo Kohn, Jeffrey L. Kolln, Eric Lent of the WI NG Museum, TSgt David H. Lipp of the ND ANG, Terry Love, Alex (A.J.) Lutz of the San Diego Aerospace Museum, Charles W. Marotske, Joseph R. McKinney of the AMARC, David Menard, Rob Mignard, Stephen H. Miller, Lionel Paul, Ned Preston of the FAA Public Affairs Office, MSgt Gregory N. Ripps of the Texas Air National Guard, Neal Schneider, L.B. Sides, Baldur Sveinsson, Norm Taylor, Ralph B. Young and SMSgt Joe Zigan of the 120th FW of the Montana Air National Guard. I'm grateful to Patti Brusk for the electronic preparation of this manuscript.

Special thanks go to my brother, Dale, who shares my enthusiasm for aviation and completed a great deal of research on my behalf. Credit is also due to the Museum of Alaska Transportation and Industry, the 148th Fighter Wing at Duluth, Minnesota, the U.S. Air Force Museum, and the U.S. Air Force History Office.

On a more personal note, I thank my wife and best friend, Deb, for her loyal support, encouragement, and understanding of the time and effort necessary to complete this book.

Every effort has been made to credit the original photographers of the images presented in this book—no easy task in view of those who have since departed, or cannot be located, and the incalculable times that material has been passed through private collections and historical agencies.

To those who flew and maintained the Deuce—and to their fallen comrades—this book is respectfully dedicated.

Introduction

To gain a better perspective of the F-102's somewhat convoluted course through Air Force history, it is best to take a look at the events in which America and her allies were engrossed since the end of World War II. In brief, the communist party roamed unfettered throughout much of Europe, many of whose nations heeled under Soviet dominance. As more countries wavered, the Marshall Plan built up steam to impede the communist advance. The Russians threatened to halt rail supplies to West Berlin, and in Asia, the menace of communist rule loomed perilously over China. Korea was divided by Russian and Western occupation forces. The oceans were no longer deemed sufficient to ensure America's safety. It was Cold War. America's efforts to stabilize the volatile quest for world domination without war required air power.

In response to the likelihood of an attack, fighter interceptor squadrons were strategically positioned, primarily across the northern edge of the United States, where they stood around-the-clock alert. Keenly aware of the responsibility of their role as America's first line of defense, the dedication of the men who flew and maintained the interceptors was matched only by their skill, confidence, and determination. They originally flew piston-powered fighters left over from World War II and first-generation jets, which were replaced by all-weather interceptors, and ultimately, the F-102. The F-102's sole mission was to intercept and destroy enemy aircraft. It was the world's first supersonic all-weather jet interceptor and the first delta-wing aircraft to become operational in the U.S. Air Force. How it garnered those attributes is quite another story.

The atomic bomb revolutionized the fundamental approach toward the development of military aircraft. The Soviet Union's mass production of bombers, whose payloads U.S. leaders were convinced would be nuclear, prompted Congress to pass the National Security Act of 1947, from which the U.S. Air Force was established. To conquer the relentless struggle for air supremacy, the Air Force in 1950 pooled all of its research and development activities into the Air Research and Development Command. It was within the ARDC that the radical weapon system concept was born. No longer were the airframe, components, and support equipment considered as separate items which came together as an aircraft, but as an air weapon system. The F-102 Delta Dagger had essentially become a mission in search of an airplane.

When the shroud of secrecy was lifted from the F-102, a remarkable system had been revealed, which the Consolidated Vultee Aircraft Corporation (Convair), Hughes, the U.S. Air Force, and the NACA nurtured to fruition. The F-102's developmental track was problematic. A then unproven theory, called the "Area Rule," literally saved the F-102. After Convair discovered that their would-be supersonic fighter could not reach the speed of sound, the area rule was applied to change the airplane's shape, allowing it to push smoothly through the sound barrier. Throughout its developmental and operational career, the F-102 was subjected to numerous modifications, resulting in 24 different production blocks, each having its own configuration.

In operational service, the Delta Dagger seldom settled anywhere long enough to become a permanent fixture. Less than three years after the type entered the Air Defense Command, the F-102 began to leave the ADC with the advent of the F-106A, which evolved directly from the F-102, and the F-101B. The "Deuce" actually went on to serve a much longer career with the Air National Guard. And scores of F-102s prolonged their usefulness in the drone program. So unsettled was the F-102's initial placement in the air defense scheme that it wasn't officially named the "Delta Dagger" until 1957, having first been dubbed the "Machete."

For the pilots, it was simply the "Deuce," and from their standpoint, it was a welcome sight and collectively hailed as a "pilot's dream." The aggressive and patriotic spirit that prevailed in the fighter pilot community left no doubt that Deuce drivers would make maximum use of the F-102 to accomplish their mission, even if that meant knocking the tail off a Russian bomber or diving through it. They guided their massive F-102s towards intercept points, often in weather that kept most prudent flyers grounded. Although imposingly large, the F-102 displayed sleek, unmistakably classic lines and was the embodiment of power, which, to the American public, translated to a sense of security. Convair exploited that image with a widely-publicized ad which depicted F-102s zooming skyward in the dawn, while a milkman watched from a doorstep—the headline read, "Freedom has a new sound."

The bombers of the Soviet Union never attacked. Whether the Russians truly believed that U.S. interceptors would always be there seems to no longer matter. However, it can be said with certainty that the F-102 was the first line of defense and is very likely why the attack never came.

Like many, my recollection of the Cold War, and specifically the F-102, revolves around a youthful fascination with airplanes, a preeminence that was nurtured at local airports, scale modeling clubs, and the Civil Air Patrol. Irreverent in our grasp of the finality of nuclear holocaust and the paranoia that swept the nation, by virtue of our youthful innocence, we were left with an attraction to the simpler aspects of military hardware, safe from its devastating implications. Growing up in the state of Wisconsin, which hosted several F-102 units, afforded regular opportunities to watch the sleek jets pierce the sky, hear in awe their thundering crescendo, and occasionally, view them close up. Gathering war clouds in Southeast Asia steeled in many of us a sense of patriotic calling, which changed forever our perception of the nature of warplanes. That ember of enthusiasm, however, coupled with a profound respect for the Delta Dagger pilot and his airplane, kindled my desire to begin work on this historical account.

Wayne Mutza
Mequon, Wisconsin
February 1999

Delta Dagger Development

The origins of technological achievements can often be traced back to their simplest forms. Delta-wing aircraft have their roots in nature; not birds, as one might expect, but the flying seed of a climbing vine found in the jungles of Indonesia. The seed's perfect aerodynamics captivated German naturalist Dr. Fredrich Ahlborn, who passed his findings to Austrian engineers Igo Etrich and Franz Wels in 1903. Enthralled by the mystery of aircraft stability, Etrich and Wels built flying scale craft, based on the flying seed, and in 1906, constructed a man-carrying glider. A powerplant was added in 1907, and by the end of World War I, Etrich had developed a popular monoplane called the "Dove." Taking the cue from those pioneers, German aerodynamicist Dr. Alexander Lippisch continued research of the delta design, named for the fourth letter of the Greek alphabet, whose symbol is the triangle. Lippisch excelled in the field of tailless aircraft and eventually began experimenting with rocket propulsion. He worked diligently on glider and piston engine-powered delta-wing designs until the end of World War II, by which time Germany had gained substantial ground in wind tunnel testing of sonic flight. Evidence of Dr. Lippisch's research and design studies first reached the United States in 1945, which set the stage for American development of delta-wing aircraft.

After pouring over the Lippisch reports in 1945, Convair deemed that the design merited further investigation and could lead to a breakthrough in wing design that offered numerous advantages. Concurrently, the U.S. Army Air Force announced its requirements in August 1945 for penetration, all-weather, and interceptor fighter aircraft. The latter appealed to Convair,

who signed a contract with the USAAF on 25 June 1946 to begin extensive testing of the delta-wing interceptor concept. Convair began work on a number of scale models and a full-scale mock-up, which the Air Force designated XP-92. Preliminary studies convinced engineers that the delta configuration's inherent stability would permit a high performance aircraft to operate in supersonic regions. They also concluded that the triangular wing would be stable and easily controlled. Dr. Lippisch had been convinced that the pitch and roll control of aircraft with large, triangular wings was possible with elevons (combination elevators and ailerons), which eliminated the need for a tailplane. The delta-wing was a relatively simple, lightweight structure, with minimum components and a large internal fuel capacity.

The handling characteristics of delta-winged aircraft differed dramatically from those of conventional aircraft. To the design's credit, its uniqueness was largely advantageous. Stalls were nonexistent. Instead, as the aircraft slowed, it began to sink until eventually it could not maintain level flight—a welcome deviation from the pitch-up associated with other types. There was no tendency to spin or lose control at extreme nose-high altitudes. In addition, the delta's low wing loading offered excellent maneuverability, especially at high altitudes. For high speeds, it was imperative that the ratio of wing thickness to wing chord be as low as possible. Since a wing could be thinned only to a certain point before it became a solid structure, designers had more flexibility in increasing the wing chord. The result was a wing root that spanned the greater part of the fuselage length and narrowed out to very small, but aerodynamically efficient, wing tips.

A radical design for the period, Convair's Model 7 (XP-92) full-scale mockup was the result of the firm's extensive delta-wing studies. Most unusual was the cockpit's location in the nose spike. (General Dynamics)

Using the same wing planform as the XP-92 mockup, the XF-92A differed significantly by its contoured fuselage and conventional cockpit location. (General Dynamics)

Protruding from the aft ventral fuselage is a support for the tail-heavy XF-92A. Since the machine was underpowered since its inception, Convair replaced the XF-92A's engine with a more powerful J-33-A-29 with afterburner during 1950. The later version is seen here with a camera housed in the fairing on the vertical fin. (General Dynamics)

The XF-92A never flew again after this accident, caused by nose gear failure, on 14 October 1953. A total of 25 NACA flights were made from the Dryden Flight Research Center, Edwards AFB, California, from 18 May 1951 to 12 March 1954. In NACA tradition, the aircraft was painted white, while trailing surfaces were left natural metal. Tufts were attached to the right wing for airflow studies. (USAF via Marty Isham Collection)

Work on the XP-92 began during late 1946, and by early 1948, the full-scale mock-up was completed. Its wing leading edge was swept back 60 degrees, and it had a triangular vertical fin. The interceptor was a mere 38 feet long and had a wing span of 31 feet, 3 inches. Its maximum design speed was estimated to be Mach 1.75 (approximately 1,165 mph) at 50,000 feet. It weighed just over five tons empty and had a takeoff weight of 29,000 pounds. Figured into the design was an armament package which comprised four 20mm cannon in an intake spike in the nose, directly beneath the cockpit.

Faced with post-war budget constraints, the USAAF approved construction of a single aircraft, which would serve as a testbed to explore the flight characteristics of the configuration. The unusually relaxed Air Force specifications dictated only that the machine, which it labeled the XP-92A, be powered by a conventional turbojet engine, be of simple construc-

One of the F-102's prime competitors in the interceptor development program was the Republic XF-103. When surpassed by the Convair F-102, the XF-103 was canceled during 1957, while still in the mockup stage. (General Dynamics)

The final version, minus the nose instrument probe and landing gear doors, as it appeared during use as a public relations vehicle. The fairing atop the exhaust section housed a drag chute, which was added in 1953. Its complete serial number was 46-682. (General Dynamics)

The XF-92A was to appear in the Howard Hughes movie "Jet Pilot," starring John Wayne. However, footage with the blue and gray craft ended up on the cutting room floor. (USAF via Marty Isham Collection)

Republic's XF-91A Thunderceptor received considerable attention when the Air Force found itself hard-pressed to acquire an interim interceptor. At the time the XF-91A was canceled, two experimental models had been built. Both were fitted with rocket engines to augment their turbojets and were used for high-speed armament tests. (Republic)

Photographed from a third chase plane, the first YF-102 prepares for landing at Edwards AFB, followed by an F-86A and F-80B. (USAF via Marty Isham Collection)

tion, and be completed in the shortest possible time. Since it also called for the use of available materials, the craft emerged as an interesting hybrid, using North American FJ-l main landing gear, a Bell P-63 nose landing gear, a Lockheed P-80 engine and hydraulic system, the exhaust section and ejection seat from a Convair XP-81, and a control column and brake parts from a Consolidated Vultee BT-13.

As construction of the XP-92A got under way, the XP-92 mock-up and associated tests suffered economic problems. The Air Force view that the craft was impractical for any use other than research, coupled with the National Advisory Committee for Aeronautics' (NACA) assertion that its disproportionate wings and fuselage precluded its status as a super-

sonic contender, were clear indications that its days were numbered.

When the XP-92A (Convair Model 7-002) was rolled out during November 1947, it differed significantly from its inanimate predecessor. Although the wing area of 425 square feet remained unchanged, the fuselage was lengthened to 42 feet, 5 inches, and the vertical fin was enlarged. The fuselage was contoured and incorporated a conventional cockpit. After extensive wind tunnel tests at NACA's Ames facility, the craft was fitted with an Allison 533-A-21 engine and shipped to California's Muroc dry lake, where it became airborne during taxi trials on 9 June 1946. Its first official flight occurred on 18 September. Contrary to what many experts predicted, it did

Northrop's F-89 Scorpion began air defense duties during June 1951. Operational problems with the F-89, which rendered it incapable of contending with Soviet bombers, placed increased pressure upon the development of the ultimate MX-1554 interceptor. (USAF)

The successor to North American's famed F-86 Sabre, the F-100 Super Sabre broke the sound barrier during its first flight in 1953. Considered during 1951 as an interim interceptor pending the ultimate F-102, the F-100 went on to become a successful tactical fighter. Pictured here is the first F-100A. (Warren M. Bodie)

This view of the first YF-102 in October 1953 clearly shows the "barrel" fuselage prior to application of the area rule. (USAF via Marty Isham Collection)

not flip over on its back or prove impossible to control. In fact, Convair chief test pilot Ellis "Sam" Shannon found that control was satisfactory.

In view of the XP-92's cancellation during 1948, the role of the XF-92A was shifted from interceptor prototype to delta-wing research platform in 1949 (it had been redesignated when the Air Force changed the "P" designation (Pursuit) to "F" (Fighter) on 11 June 1948. As such, it was instrumental in Convair's success in developing the F-102, F-106, XF2Y-1 (the world's first delta-wing seaplane), and the legendary B-58 supersonic bomber.

Ventures into the virgin supersonic regions had taken place on other fronts. On 14 October 1947, less than one

Photographed at Edwards AFB on 16 September 1954, the second YF-102 displays its distinctive profile following major nose and tail changes. (General Dynamics)

This view of the second YF-102 outlines the ample dimensions of the missile bay. (USAF)

Seen here is the original location of the speed brakes, which were relocated aft of the rudder when the area rule fairings were added to the aft fuselage. Visible in the background are jets common to the period: F-84, F-86, and F-89. (General Dynamics)

The "straight leg" nickname given the early landing gear configuration is apparent in this view of the second YF-102. (Leo Kohn)

In this comparison, it becomes apparent that the YF-102's cockpit was set farther back in the fuselage, resulting in a nose much longer than that on later models. (USAF)

month after the U.S. Air Force was established, the Bell X-I had slipped through the sound barrier. During 1949, North American began transforming its famed F-86 Sabre into the F-100, which would go supersonic on its first flight in 1953.

During mid May 1949, the XF-92A was turned over to the Air Force for Phase II testing. Major Frank "Pete" Everest and Captain Charles "Chuck" Yeager completed the trials by year's end. Underpowered from the beginning, the aircraft's Allison J33-A-23 (which supplanted the original -21) was replaced by a J33-A-29 with afterburner that offered more than 3,000 pounds more thrust, for a total of 7,500 pounds. Al-

though Mach 0.95 was achieved, the -29 proved unreliable, necessitating numerous engine changes. After a more powerful J33-A-16 was installed, test pilot Scott Crossfield flew a series of tests, during which violent pitch-ups occurred during turns. Combinations of wing fences were tried, which eventually solved the problem.

A test flight that ended with nose gear failure on 14 October 1953, plus the impending arrival of the YF-102 in 1954, spelled the end of the XF-92A. It was removed from the inventory and used as a public relations prop until it was placed on display at the University of the South at Sewanee, Ten-

XP-92

AIRPLANE   MOCK-UP

CONSOLIDATED VULTEE AIRCRAFT CORPORATION · SAN DIEGO · CALIFORNIA

The third YF-102 lands at Holloman AFB, New Mexico, with its missile bay doors open. (USAF)

Convair added a touch of humor to the complexities involved with developing the XP-92 with this drawing, used for the cover of the aircraft's manual.

A l/10th scale XF-102 jet "exit" model in a 6 by 8-foot wind tunnel during late August 1953. (General Dynamics)

The sixth YF-102 wore the Air Force Flight Test Center emblem on the fuselage, indicating its long term assignment to the facility. (W.J. Balogh, Sr. via David Menard)

nessee. In 1969, it was taken to the U.S. Air Force Museum, where restoration finally began during 1997. During its test period, Air Force, Convair, and NACA pilots flew a total of 62 hours during 118 flights.

Only two years after World War II had ended, America and her allies faced the reality of Cold War. Just weeks before the U.S. Air Force was formed, Russian intercontinental bomber fleets filled the camera lenses of allied intelligence agents, giving rise to a genuine fear that North America was vulnerable to air attack. Fortifying that perception, while eroding the Free World's sense of security, was the detonation of the Soviet Union's first nuclear device on 29 August 1949. U.S. military leaders wasted no time in their quest for a high performance interceptor that could exceed the speed and altitude of Soviet bombers. Warnings from intelligence sources and the limited growth potential of the U.S. jet interceptor fleet prompted a call to action.

On 13 January 1949, the Air Force initiated an Advanced Development Objective, which called for an advanced, one-of-a-kind interceptor to deal with the burgeoning Soviet bomber fleet. The design was labeled the "1954 Interceptor" for the year it was expected to become operational. Recog-

Various combinations of wing fences were tried on early models, with the final configuration seen here on the fifth YF-102 in August 1955. The "F" in the buzz number stood for "Fighter," while the "C" was the designator for the F-102. (McDonnell Douglas Corp.)

A YF-102 at the NACA High-Speed Flight Station, Edwards AFB in 1955. Wing fences were attached by NACA personnel to alleviate pitch-up encountered during test maneuvers. This was the first of the series assigned to the NACA. Others were eventually rotated through various NACA/NASA facilities for a wide variety of uses. (NASA-Dryden Flight Research Center)

nizing that complex weapons systems could no longer be simply "added on" to a completed aircraft, the Air Force directed that the interceptor be developed under the Weapon

NACA engineers Cliff Morris and Thomas Sisk perform a calibration on an Alpha-Beta Vane, which measured angle-of-attack and sideslip during flight maneuvers. This view of the YF-102 (s/n 53-1785) on the NACA ramp at South Base, Edwards AFB, shows to good effect the cockpit framing and early two-piece landing gear configuration. (NASA-Dryden Flight Research Center)

System Concept. The thrust of the untried concept was to integrate systems at the onset, ensuring total compatibility in the final product.

The 1954 Interceptor project was officially designated WS-201A, which called for a weapons system comprising primarily air-to-air missiles controlled by an all-weather search and fire control radar, embodied in an airframe capable of supersonic speeds.

While aerodynamic and structural engineers worked to refine the delta-wing, the Air Force, shortly after its inception, became partners with the budding Hughes Aircraft Company. Together, they explored unique ways to intercept and destroy enemy aircraft, at night and in bad weather. Firms that were well-established in radar development, such as Martin, North American, RCA, GE, and Westinghouse, also conducted research, however, it was the aggressive team of Hughes engineers that showed the most promise.

The accepted system was called "collision-course," a form of which had already been in use aboard F-86D, F-89, and F-94 interceptors. The technique had the aircraft's radar linked to the flight controls to direct the interceptor abeam of a target's track at the precise time their paths were about to cross. When the opponent's flank profile filled the interceptor's radar gunsight, a dense volley of rockets would bring it down. The novel method allowed faster interception, ruling out the chance of losing a faster aircraft in a traditional chase from behind.

The first requirement of the WS-201A system to be addressed concerned the armament and electronic fire control segment, which was designated Project MX-1179. On 18 June 1950, the Air Force announced that it was seeking proposals for an airframe structurally capable of speeds exceeding Mach 1 (760 mph) and altitudes above 50,000 feet. By the time bidding closed in January 1951, nine proposals by six aircraft manufacturers had been submitted. Republic submitted three

The second of four YF-102As after retrofitting of a new canopy. Unusual is the fact that the exhaust section was included in its fresh paint job. (Leo Kohn)

Apparent in this view of the first YF-102A are the many improvements over the original YF-102 model. (USAF)

bids, North American two, and Chance-Vought, Lockheed, Douglas, and Convair each presented single offers. On 2 July, the Air Force announced that it had selected Republic, Lockheed, and Convair to proceed with development of their designs through the mock-up stage. Deeming it unwise to finance three concurrent development programs, the Air Force soon canceled Lockheed's entry.

Meanwhile, Hughes was awarded the contract for the electronics control system (ECS) during October 1950. Hughes drew from their experience with their E-1 fire-control system (installed in the F-86 and F-94 interceptors) to begin work on the MA-1, which was slated for production in 1953.

After extensive evaluation of design proposals by the three firms that remained in the airframe competition, Convair was awarded the contract on 11 September 1951 for what was

termed the MX-1554 airframe. The same agreement authorized the installation of the Westinghouse J-40 turbojet, pending availability of the more powerful Wright J-67. Convair's entry, combined with the J-67 and Hughes ECS, was officially designated the F-102. Having evolved from the XF-92A, which by that time was nearing Mach 1, Convair's delta-winged design was expected to reach Mach 1.93 at 62,000 feet.

Although Republic's primary proposal was edged out, its compelling, futuristic design intrigued the Air Force to the extent that its development contract was extended. Designated the XF-103, the all-titanium machine was intended to fly at Mach 3 at an altitude of 80,000 feet, an incredible achievement for the period. In the ensuing years, however, the program suffered a number of setbacks, which resulted in its termination in September 1957.

The first of four YF-102As and the first to reach Mach 1 in level flight. Obvious changes over its predecessors are the redesigned canopy, wraparound outboard wing fences, redesigned air intakes, upturned wing tips, and a longer nose. (USAF)

The fifth YF-102 is prepared for "drop tests" at the Convair facility. (General Dynamics)

This YF-102 suffered nose gear failure during landing at Edwards in September 1954. (General Dynamics)

Production of the "Ultimate Interceptor" was scheduled for 1953 or early 1954. Convair was painfully aware of the challenges it faced in filling the tall order, especially since their delta-wing XF-92A had yet to reach Mach 1. Added to that was the apprehension of designing an internal missile bay for a new air-to-air missile that had yet to be built.

Anxious for their Ultimate Interceptor, the Air Force expedited the F-102 program in January 1952 by implementing the Cook-Craigie Plan, named for its architects, Generals Orval R. Cook and Laurence C. Craigie. The objective of the novel, untried plan was to begin slow-moving production that bypassed the X-designated prototype and focus on service test versions. The program paid dividends in time saved by making it possible to quickly incorporate changes brought about by tests into an active production line. While the small number of aircraft were built to fulfill test requirements, tooling and preparations were made for full-scale production. Aircraft delivered early in the program could be rerouted through the system to be brought up to the latest standards. The Cook-Craigie Plan was high risk and worth putting into effect only if there was a high degree of certainty that full-scale production would ensue. Confident in the design, the Air Force took that risk in view of the then-active XF-92A.

By December 1951, flaws in the multi-faceted MX-1554 concept had become apparent. The J-67 powerplant and the MA-1 fire-control system would not be ready for the final systems merge, forcing the Air Force to change its plans. That prompted the decision to proceed with an interim version of the F-102 as work continued on the Ultimate MX-1554 Interceptor. The temporary aircraft would be called the F-102A,

Convair test pilot John Fitzpatrick climbs aboard a YF-102 prior to a test flight during spring 1954. Clearly shown is the early canopy structure and the sweeping angle of the air intake configuration. (General Dynamics)

This view of the YF-102 emphasizes the immense fuselage prior to application of the area rule. (General Dynamics)

Dressed in an Air Force T-1 pressure suit, Convair chief F-102 test pilot Richard L. Johnson prepares to take the first YF-102A aloft from Edwards AFB. Rubber tubes along the suit's arms and legs were inflated during high altitude flight to tighten laces and apply pressure, protecting him in case of cockpit pressure failure. The emergency bottle on his left leg provided oxygen in the event of bailout above 15,000 feet. Visible in the background is the rare XA-45, later dubbed the XB-51. (General Dynamics)

while the fully-developed version would become the F-102B. By 1956, engineering changes had widened the differences between the two so extensively that the F-102B received its own designation, the F-106.

The Air Force had strongly considered other aircraft then under development as interim interceptors. They were Republic's XF-91A "Thunderceptor," the Douglas XF4D-1 "Skyray," and North American's "Sabre 45," later known as the F-100A. On 7 November 1951, the decision was made to drop the XF-91 and continue development of the F-100 as a tactical fighter. The Skyray went to the Navy.

Confidence in the design remained so high that the Air Force ordered two prototypes in January 1952, seven production F-102s in February, and a static test platform in August. The manufacture of the preliminary order began during April 1952. Despite the engine and fire-control system delays, the Air Force increased the number of production ma-

chines in September 1952 from 7 to 40. In view of the war in Korea and the growing Soviet bomber threat, the Air Force made a total of 21 amendments to the original contract to accelerate the program. Although a great deal of emphasis was placed on the development of the service test aircraft

Modified with the Case XX wing, this F-102 was one of two aircraft used by the AFFTC in mid-1957 to evaluate the concept. The missile bay was gutted to make room for test equipment and ballast was added to simulate a full weapons load. (Larry Davis Collection)

The third-built F-106A is compared to the first F-102A modified with the Case X wing in 1958 at Edwards AFB. (USAF)

Seen here on its first flight, the first YF-102A shows off the form that enabled it to fly supersonic in level flight. Clearly evident are the cambered leading edges, reflex wing tips, and boundary layer wing fences, which improved handling at lower speeds. The needle nose projection is the pitot tube to measure air pressure, which, in turn, provided airspeed readings. Also obvious is the sharper-angled canopy, with flat windshield and slightly bulged side panels. (General Dynamics)

with the J-40 or J-57 engine and E-4 electronic control system, it remained the primary goal to produce the ultimate F-102A with the Wright J-67 and Hughes MX-1179.

When it was discovered that the J-40 engine lacked the thrust to fly the F-102 at peak performance, the switch was made to the Pratt & Whitney J-57 turbojet, scheduled to enter production in early 1953. The J-57 was the world's first jet powerplant to develop 10,000 pounds thrust. Originally designed for the B-52 bomber, the J-57 powered the YF-100A when it became the world's first fighter to fly supersonic in level flight.

Also suffering development delays was the MX-1179 fire-control system, which would not become available prior to late 1955. A modified Hughes E-4, called the E-9, was used instead. It was renamed the MG-3 when its ability was substantially improved, and was eventually upgraded, becoming a permanent feature in the F-102. More detrimental to the F-102's development were problems encountered during NACA wind tunnel tests in early 1953. It became apparent that initial performance figures were overly optimistic and that sufficient allowances had not been made for the aircraft's aerodynamic drag. The F-102 would fall short of its goal to reach

The dissimilarity between the second production F-102A and later models shows not only that limited production took place during flight testing, but the large number of modifications instituted over the years. Though upgraded with the area-ruled fuselage, new canopy, and redesigned air intakes, the aircraft retained the original short tail fin and nose. (General Dynamics)

These views illustrate the vast differences between the YF-102 and F-102A. Finished in the NACA white livery, serial number 54-1374 is seen while assigned to the NACA High-Speed Flight Station at Edwards in 1956. Evident is the area-ruled fuselage taper in relation to wing span, ending in a widened aft fuselage. The broad span of the elevons is also apparent. (General Dynamics)

Mach 1. Convair's team was left with no alternative but to redesign a new airframe that conformed to the "area rule" to drastically reduce transonic drag. The area rule was discovered during 1951 by Richard T. Whitcomb of NACA's Langley Aeronautical Laboratory and later verified by rocket-powered model tests at NACA's Pilotless Aircraft Research Station at Wallops Island, Virginia. When transonic tests at NACA's wind tunnel showed that the "drag hump" at sonic speed was greater than the aircraft's capability, Convair and NACA engineers agreed that salvation lay in the area rule. The revolutionary concept, though kept under tight military security, was discreetly made available to the aircraft industry in 1952. Grumman benefited from the concept when applied to its F11F-1 "Tiger," as did Chance-Vought in the development of its F8U "Crusader."

Like the delta-wing, the area rule had its origins in Germany's wind tunnels during World War II. The German aerodynamicist Kuchemann conducted flow studies over the wing root of a swept-back wing and fuselage combination. He found that the flow turned in toward the fuselage and out again, requiring that the fuselage be contoured to match the flow. U.S. intelligence teams obtained his data, which was called the "Kuchemann Coke Bottle." Whitcomb's discovery was given the same name, which was a misnomer since his area rule applied to general shapes and alluded to a different outcome. He reasoned that interference drag, which was responsible for the greater portion of transonic drag rise, was caused by the interaction of aerodynamic components and depended almost entirely on the distribution of the aircraft's total cross-sectional area along the direction of flight. By indenting the fuselage, thereby reducing its cross-section over the wing, drag was greatly reduced.

It was too late to incorporate the vast changes associated with the area rule into the YF-102 and early production aircraft since that contract had been finalized on 12 June 1953 and production had begun. It was, however, a stage in the F-

102 program where the Cook-Craigie plan proved its worth. Production under previous concepts would have resulted in hundreds of aircraft that required modification, versus the initial batch built at a slower pace.

The YF-102 was basically an enlarged XF-92A with a longer, pointed nose and lateral air intakes. It was to carry six Hughes Falcon air-to-air missiles in a bay that occupied the greater portion of the lower fuselage. As backup, 24 2-inch or 2.75-inch rockets were to be carried within the missile bay doors.

Following completion, the first YF-102 lifted off from Edwards on its maiden flight on 24 October 1953, with Convair test pilot Richard C. Johnson at the controls. Power was derived from a J57-P-11, rated at 10,900 pounds thrust and 14,500 pounds with afterburner. Severe buffeting occurred at Mach 0.9 and, as had been foretold by wind tunnel tests, Mach 1 was beyond reach. A partial engine failure during its seventh flight, which took place on 2 November, caused the aircraft to crash at the end of the runway. While Johnson recovered from serious injuries, flight tests resumed with the second machine's first flight on 11 January 1954, flown by Sam Shannon. Its performance also confirmed that the design was drag-limited to Mach 0.98, with a 48,000-foot ceiling. In fact, the F-86D that the F-102 was to replace had a higher ceiling and nearly the same speed. Modifications that included cambered wing leading edges, upturned (reflex) wing tips, a tapered fuselage with 7-foot extension, and repositioned wings and vertical fin not only added 3,500 pounds, but failed to improve performance. The F-102 program was in trouble, and the time had come to put the area rule into play.

Mindful of undercurrents in the Air Force that the program might be canceled, Convair, after finally accepting the necessity for a drastic redesign, worked feverishly to come up with a new package.

In just under four months, the first of four YF-102As was rolled out at the San Diego facility. Dubbed the "Hot Rod" for its anticipated performance and the speed at which it was built, the aircraft had been pulled from a batch of 12 aircraft in various production stages. In keeping with the area rule, its fuselage was narrowed, forming what became known as the "NACA Ideal Body," or more popular, "Marilyn Monroe" design.

The multi-faceted area rule principle dictated that not only the fuselage be pinched, but that other alterations be made, as well. A perfect example of the area rule's dictum that "it was not always necessary to remove area, but also add area," was the enlarged aft fuselage. Bulged, streamlined areas added to the F-102 fairings that extended beyond the exhaust filled in the area diagram to delay flow breakaway and complete the ideal shape. Similarly, the nose was lengthened an additional four feet to increase the fineness ratio (length to diameter), minimize mach cone interference with the wing, and provide ample space for radar components. In addition, the cockpit was moved forward and covered with a new canopy. The engine intake ducts were modified, and the wing camber was reworked. Especially important was the change to an interim J-57-41 engine, pending the arrival of the J-57-P-23. The -41 was basically an -11 modified with bleed air probes to eliminate fumes in the cockpit. An added benefit was the significant weight reduction that resulted from the use of a lighter engine support structure. Armament consisted of six air-to-air missiles, augmented by 24 rockets launched from tubes built into the missile bay doors.

The YF-102A made its first flight on 20 December 1954, and the following day, easily reached Mach 1.2 and an altitude of 53,000 feet.

Meanwhile, the eight additional straight-fuselage YF-102s were completed while the redesign was in progress. Throughout 1954 and '55, the 14 original aircraft underwent innumerable tests, during which as many changes were made both internally and externally. Powered by the J-57-P-23 engine, the first production F-102A flew for the first time on 24 June 1955. It was accepted by the Air force days later. Although a

The fourth-built F-102A (s/n 53-1794) is transported by truck at Edwards. Following close behind under wraps appears to be a Chance Vought F8U Crusader, which was figuratively kept under wraps in view of its innovative area-ruled, high wing design. (Marty Isham Collection)

marked quantity production buildup followed, a year of tests would pass before the type became operational.

The F-102 test program went far beyond any ever scheduled for a new Air Force aircraft. Ultimately, nearly 50 aircraft, including the YF-102, YF-102A, more than 30 F-102A, and four TF-102A models would be channeled into a test program that deviated significantly from the customary evaluation of one or two prototypes. In the area of aerodynamic flight loads alone, the F-102 was more exhaustively tested than any other military aircraft. The seven-phase test program was carried out simultaneously by the U.S. Air Force, contractors, and the NACA.

The largest number of aircraft were assigned to the Air Force Flight Test Center at Edwards AFB, California. Seven were put through their paces by the Air Force for specifications compliance, while another seven, including a TF-102A, underwent similar evaluation by Convair at Edwards. A total of 12 operated from Florida's Eglin AFB: ten for operational suitability trials with the Air Proving Ground and two with the Air Force Armament Center. Six F-102As and a TF-102A were used by the Holloman Air Development Center of the Air Research and Development Center at Holloman AFB, New Mexico, for weapons systems tests. The Wright Air Development Center at Wright-Patterson AFB, Ohio, used two aircraft to evaluate all-weather capability, and three went to the Cambridge Air Research Center at Hanscom AFB, Massachusetts, in conjunction with SAGE system tests. One aircraft was sent to the Air Training Command at Lowry AFB, Colorado, for training fire control technicians, while another helped with the development of inspection and repair programs at the San Antonio Air Material Area at Kelly AFB, Texas. A pair of F-102s was turned over to the Hughes Aircraft Company at Culver City, California, for fire control and automatic flight testing.

F-102s committed to test programs were commonly designated JF-102As, the "J" used to identify temporary special test status beginning in 1956. A number of F-102s diverted to special test projects appeared periodically on the Air Force inventory labeled as such.

Not only were modifications to the F-102 completed as a result of the intense test program, the type was continually upgraded throughout its service career, even after many had begun to leave the regular Air Force. Modifications were often accomplished during scheduled overhauls, called "IRAN," for "Inspect and Repair as Necessary."

One of the first major design changes resulted from an F-100 crash during October 1954 that killed its test pilot. Investigation revealed a previously unknown phenomenon associated with emerging aircraft that featured long fuselages, short wings, and supersonic conditions. Called "inertia coupling" or "roll coupling," the condition caused the aircraft to diverge violently in pitch and yaw during roll maneuvers. The solution took the form of an enlarged vertical fin, which was chosen over a stability augmentation system for its reliability. The fin was heightened 33 inches, increased in area from 68 to 95 square feet, and the forward cant of the trailing edge was changed to vertical. The new tail was test flown on the 23rd production aircraft during December 1955. It was introduced to the production process in 1956 with the 66th F-102A, while those previously built were restricted to 90 degrees of roll until retrofit.

Other major changes completed at that time included a 40 percent increase in the size of the speed brakes. When activated, the new brakes sent a signal to the automatic flight control system to momentarily raise the elevons, correcting the tendency for the nose to drop when the brakes opened. Ramps were added to the air intakes, which were later extended to eliminate high frequency noise in the cockpit and dampen buffeting.

In the interest of logistics, the Air Force decided in mid-1956 that only the 2.75-inch rockets would be used to augment the Falcon missiles. The missile bay doors of operational aircraft were field-modified to accept only the more common 2.75-inch rocket over the original T-214 2-inch model. Late in the F-102's career, rockets were deleted from the armament package.

As F-102 operational parameters were expanded, problems revealed themselves, particularly those of a consistent nature. Since its inception, the F-102 had been plagued by landing gear failures, which ceased after new serviceable struts were placed in service. Convair also came up with a solution for in-flight speed brake failures, which provided some anxious moments for the pilot when only one of the two brakes opened. As part of what was termed the Test-to-Tactical modernization program in 1957, the oxygen system was changed from a gas to a liquid system. Other changes that year included the installation of provisions for mounting external fuel tanks. The jettisonable tanks had a 215-gallon capacity, which made two-hour missions routine, but restricted the F-102 to subsonic speeds. As F-102s arrived from various test programs, -41 engines were replaced with J57-P-23s, improved afterburners replaced older types, and the standard gray finish was applied.

Foremost among the internal changes made in the F-102 during that period was update of the MG-3 radar fire control system to the MG-10. Much of the system occupied the ample nose space, while the remainder was located behind the cockpit. The MG-10 translated digital signals to analog for display on the radar scope, thereby eliminating the need for ground control for that function. Locked on to the target, the MG-10 essentially flew the interceptor heading and carried out the attack using automatic infrared or radar search. The system proved beneficial during instrument landing approaches and was further improved during 1965 with data link, which provided flight data to the pilot electronically and allowed the aircraft to be flown remotely from SAGE (Semi-Automatic Ground Equipment). As with the F-86D, the F-102 pilot was also responsible for navigation and radar intercept, which made for a busy cockpit. Prior to deployment to Alaska and Europe in 1957 and 1959, some F-102s had TACAN (Tac-

During its troublesome early period, a Deuce poses with its stablemates from the original Century Series. From front to rear are the F-104, F-100, F-102, and F-101. The latter, like the Deuce, endured a problematic start. The Voodoo first flew on 29 September 1954 and surpassed F-102 performance. (USAF)

tical Air Navigation) gear installed for compatibility with ground stations in those regions.

Also added during modernization phases was a more advanced fire control system, which was more closely integrated with the radar and fire control systems. Other problems were addressed as they proved detrimental to the F-102's operational capabilities. Fires that occurred in some F-102s during starting were traced to the air-driven starter, which gave way to a combustion type.

In the summer of 1957, during Phase IV of the test cycle, the effects on performance by a change in wing camber were evaluated by the AFFTC. The program involved two production F-102s: one with the standard "Case X" wing (s/n 56-1000) and another (s/n 56-1317) with a more highly cambered wing, called the "Case XX." Instead of the upturned tips familiar to the Case X wing, the droop of the Case XX wings continued to the extreme tip. The elevon area was increased from 67 to 70 square feet, which resulted in a lower minimum takeoff speed and improved acceleration immediately following takeoff. After nearly 17 hours of flight test between the two aircraft, the Case XX was found to have extended the maximum combat ceiling to 56,000 feet. Handling qualities were markedly improved at low airspeeds, and buffeting was diminished during power approaches. The Case XX wing also had a remarkably positive effect on climb performance and lateral control in stall approaches.

As part of the Case XX wing studies, the main landing gear of both test aircraft was slanted forward to measure its

Convair workers put the final touches on the TF-102A mockup prior to its official inspection in January 1954. Only the forward section was built since the remainder of the aircraft was identical to the single-seat version. The unusual engine intakes and cockpit style never made it to production models. Note that Convair used the designation "TF-102-A." (General Dynamics)

The TF-102A mockup cockpit. The instructor pilot was positioned on the right. Radar scopes are incorporated into the instrument panel hood. (USAF)

The first TF-102A lands at Edwards AFB during initial tests. (General Dynamics)

effectiveness, in relation to both old and new elevon sizes, in lowering takeoff speeds. The original "straight leg" gear created a greater nose-down moment, requiring a higher speed to make the elevons effective in getting the nose airborne. Since the wing and landing gear alteration was introduced late in production—during October 1957, when more than 500 aircraft had been built—earlier models were not retrofit.

Under Project Big Eight in 1963, the rockets were deleted in lieu of modifications made to accommodate combinations of advanced Falcon missiles. Since some were infrared-guided, an infrared search and track system was mounted in a distinctive, transparent dome immediately forward of the windshield. The F-102 pioneered the use of both radar and IR-guided systems to search, locate, lock-on, and attack targets in any weather, making it an extraordinarily lethal weapons system. Not surprisingly, the Soviet Union mirrored the technique.

Compared to the F-102A, the Tub's air intakes were positioned lower and molded more into the fuselage. (Author)

THE TUB
Convair's two-seat trainer arose from the need for a specialized version of the F-102A to teach pilots the unique handling characteristics of delta-wing aircraft. Air Defense Command and Air Training Command officials recognized that need early in the Delta Dagger's development, mindful of shortcomings

Its drag chute fully deployed, the first "Tub's" nose settles to the ground during tests. (General Dynamics)

This view of the TF-102A leaves little doubt as to why the aircraft was called the "Tub." The bulbous configuration of the forward fuselage imposed limits to subsonic speeds. The absence of vortex generators and the instruments attached to the pitot tube identify this as an early test aircraft. (Marty Isham Collection)

Like the single-seat variant, the trainer's radar and fire control system was contained in the nose radome. (Terry Love)

side-by-side seating permitted the interaction necessary to effectively teach pilots flying technique and radar interception. Conversely, the design was a compromise, since changes made to the refined fuselage were expected to hinder performance.

The Air Force authorized production of the trainer variant on 16 September 1953, however, problems with the YF-102 brought the move to a standstill. Finally, during July 1954, an initial order for 20 TF-102As was placed, with the first delivery due one year later. The mock-up was inspected in September 1954, and in early 1955, in view of the YF-102A's successful flight evaluations, 28 additional trainers were ordered.

The first TF-102A (Convair Model 8-12) flew from Edwards AFB on 8 November 1955, with Convair test pilot Richard L. Johnson at the controls. One month later, the Air Force awarded Convair a contract for 150 trainers, which were slated for delivery from March to December 1957.

in the T-33s and radar-equipped B-25s used for pilot transition to the F-86D and F-94. As a Combat Proficiency Trainer, the TF-102A, best known as "The Tub," was a dramatic deviation from the conventional tandem-seat configuration. Its

With its forward fuselage, outer wings, and most of the vertical fin painted fluorescent red-orange, this TF-102A is seen during tests at the AFFTC, Edwards AFB during December 1957. (USAF)

Fully modified TF-102A of the Air Defense Weapons Center. (USAF)

In designing the TF-102A, the length and contours of the original F-102A's forward fuselage underwent a radical change to form the side-by-side cockpit arrangement. The nose section was sculpted in accordance with the area rule principle, and from the cockpit rearward, the trainer and single-seat interceptor were identical. The Tub's broad frontal cross-section had the predictable effect on performance, however, supersonic flight was possible in a shallow three to five-degree dive. Although the trainer was not beset by problems like early single-seat variants, it endured its share of difficulties.

When test pilot John M. Fitzpatrick took the TF-102A aloft to begin flight tests, he experienced severe buffeting and ter-

The first TF-102A at Chanute AFB, Illinois, wearing the emblem of the Air Training Command on the tail fin. (David Menard)

The enlarged vertical tail and canopy vortex generators were introduced on this TF-102A, the third in production. It is seen here near Edwards AFB on 3 January 1958 while assigned to the AFFTC. (USAF)

A Tub of the 4780th Air Defense Wing at Perrin AFB, Texas. TF-102As were a staple at the Air Defense Command training base. (John Guillen Collection)

This view of the TF-102A preserved at the Pima Air & Space Museum shows how the bulbous cockpit blended into the streamlined fuselage. (Dale Mutza)

minated the flight at Mach 0.8. Photo coverage of the tuft-covered aircraft on the following flight revealed that flow separation over the bulbous canopy was the culprit. Reducing the height of the canopy proved effective, however, it prevented the pilot from seeing the runway during the critical landing flare. Various types of canopy extensions were tried. Some helped, some made it worse, but none erased the problem. Finally, the right arrangement of vortex generators attached to a strengthened canopy frame eliminated the vexation.

The only other significant concern noted by TF-102A pilots was the cockpit layout. Although the student pilot's position was a facsimile of the F-102A's, the instructor pilot's throttle was opposite, on the right side of the cockpit. That meant that pilots flying right seat had to depart from their practice of left hand on the throttle and right hand on the control stick.

The TF-102A's buffeting problem prompted the Air Force to halt production. The hold was rescinded during June 1956

Following active Air Force service, TF-102As filled the ranks of Air National Guard squadrons. This fully modified example served Hawaii's 199th FIS. (Lionel Paul)

TF-102A of the 4780th ADW at Perrin AFB. (Jerry Geer)

This Tub, seen at San Diego during May 1957, displays the original practice of applying titles and buzz numbers along the full length of the fuselage. (Merle Olmsted)

A TF-102A of the California Air National Guard. (Ron Montgomery)

During 1967, this was the last of 87 Deuces to receive air-to-air refueling modifications. It was then flown to Clark AB as a 509th FIS replacement aircraft. (S. Kraus via Norm Taylor)

after successful testing of the third TF-102A, which introduced the canopy fix and enlarged vertical tail. During the same period, the Air Force trimmed its orders for trainers and canceled the last 87 machines. Despite the reduction, deliveries still lagged six months behind schedule.

Upgrades in TF-102As paralleled those introduced to the line of single-seaters. As a result, the total number of TF-102As were divided almost evenly between Case X and Case XX wing configurations. Although the Tub was fully capable of operating as a tactical interceptor with the basic armament package, the Hughes MG-10 fire-control system was seldom installed.

Besides pilot training, TF-102As were involved in a number of noteworthy programs. One example was used for evaluation of North American's automatic approach and landing system during 1961 and '62. Other Tubs were flown by pilots entering the Convair B-58 bomber program during the early

While assigned to the 4750th Test Squadron, this TF-102A (s/n 56-2345) wore photo registration marks. The full-length elevons are clearly seen. (L.B. Sides Collection)

Col. Leon Gray and then-Senator Barry Goldwater prior to a flight in a TF-102A of the 498th FIS. In keeping with the unit's "Geiger Tiger" motif, the pilots' helmets were painted accordingly. (USAF)

The inlet ramp, or separator plate, was an early modification of the engine intake called the "buzz fix," since its purpose was to eliminate high-frequency buzzing of the fuselage side panels during flight. (Author)

Seen here near the end of F-102A production, 16 Delta Daggers (s/n 57-0851 to 0866) await final assembly. Convair's innovative production methods eased the complexities of mating various assemblies. (General Dynamics)

1960s. Prospective "Hustler" drivers spent six weeks at Perrin AFB, Texas, for intensive instrument flight training in the T-33 and TF-102A. During early 1965, four TF-102As were committed to special test programs and subsequently redesignated NTF-102As.

When the TF-102A line was shut down in mid-1958, a total of 111 examples had been produced, only two-thirds the number originally programmed.

THE AIRPLANE

As the first aircraft to be ordered by the U.S. Air Force under the Cook-Craigie production plan, the F-102 became known in manufacturing circles as a "Production man's dream." To build the Delta Dagger, Convair transferred their chief tool engineer at the Fort Worth plant, A.P. Higgins, to San Diego Plant 2. His idea was to break down the airplane into subassemblies small enough for easy handling, yet large enough to utilize maximum manpower. A harvest of new ideas guided F-102 production, which combined conventional mechanized, heavy press, and half-shell construction methods for the first time.

During the dawning of the jet age, American designers subscribed to the theory that wings should be as thin as possible. In reality, that gave the F-102's wing a thickness ratio of only five percent. It was so thin that elevon controls, which could not be contained within the wing, had to be housed in external fairings. The basic wing structure consisted of two fuel cells in each wing: one forward and the other rearward of the landing gear well. Wing leading edges were made of stainless steel through which hot air was ducted for anti-icing. Main landing gear struts and doors retracted into the wing, while the bulkier wheel assemblies recessed into the fuselage. The variety of wing fence combinations, which alleviated pitch-up tendency and improved low-speed control, culminated in a

The final version of the speed brake, which was changed in shape and increased in area over earlier configurations. (Author)

Positioned just aft of the tail hook, dual data link antennas began to appear on F-102s in mid-1965.

pattern of one outboard, which wrapped around the leading edge, and a smaller inboard fence. The F-102's wings spanned 38 feet, 11 inches, and its root chord was 29 feet, 9 inches. With the exception of the elevons, dimensions remained the same when the Case X wing was replaced in favor of the Case XX. The elevon span then increased from

12 feet, 11 inches to 14 feet, resulting in an area increase of just over 3 square feet.

To compensate for the F-102's thin wing, an immense fuselage was necessary to contain all major components. The fuselage was built of conventional frames and stringers, which were riveted and Scotchwelded, the latter a technique for

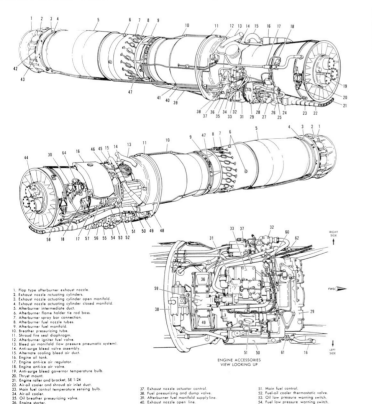

1. Flap type afterburner exhaust nozzle.
2. Exhaust nozzle actuating cylinders.
3. Exhaust nozzle actuating cylinder open manifold.
4. Exhaust nozzle actuating cylinder closed manifold.
5. Afterburner intermediate duct.
6. Afterburner flame holder tie rod boss.
7. Afterburner spray bar connection.
8. Afterburner fuel nozzle tubes.
9. Afterburner fuel manifold.
10. Breather pressurizing tube.
11. Shroud fire seal diaphragm.
12. Afterburner igniter fuel valve.
13. Bleed air manifold (low pressure pneumatic system).
14. Anti-surge bleed valve assembly.
15. Alternate cooling bleed air duct.
16. Engine oil tank.
17. Engine anti-ice air regulator.
18. Engine anti-ice air valve.
19. Anti-surge bleed governor temperature bulb.
20. Thrust mount.
21. Engine roller and bracket, SE 1-24
22. Air-oil cooler and shroud air inlet duct.
23. Main fuel control temperature sensing bulb.
24. Air-oil cooler.
25. Oil breather pressurizing valve.
26. Engine starter.
27. Primary hydraulic pump.
28. Engine junction box.
29. Oil pump and accessory drive housing.
30. Constant speed drive unit oil filter.
31. Fuel pump transfer valve.
32. Afterburner fuel control.
33. Engine and afterburner fuel pump.
34. Fuel flowmeter.
35. Engine and afterburner fuel supply inlet.
36. Right-hand ignition transformer.
37. Exhaust nozzle actuator control.
38. Fuel pressurizing and dump valve.
39. Afterburner fuel manifold supply line.
40. Exhaust nozzle open line.
41. Exhaust nozzle closed line.
42. Exhaust nozzle actuating cylinder closed line
43. Exhaust nozzle actuating cylinder open line.
44. Anti-surge bleed governor.
45. Fuel tank pressurization line fitting.
46. Alternate cooling air valve.
47. Aft engine mount.
48. Shroud cooling air duct check valve.
49. Left-hand ignition transformer.
50. Fuel-oil cooler.
51. Main fuel control.
52. Fuel-oil cooler thermostatic valve.
53. Oil low pressure warning switch.
54. Fuel low pressure warning switch.
55. Constant speed drive unit gear box.
56. Secondary hydraulic pump.
57. Oil tank drain valve.
58. Left-hand forward engine mount.
59. Accessory drive adapter.
60. Exhaust nozzle control fuel filter.
61. Main oil strainer.
62. Igniter fuel valve fuel filter.
63. Deleted.
64. Constant speed drive unit oil recirculating valve.

ENGINE ACCESSORIES
VIEW LOOKING UP

DETAILS OF THE PRATT & WHITNEY J57

Details of the nose landing gear and door. (Lennart Lundh)

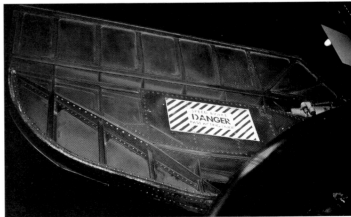

Details of the interior of the left speed brake. (Author)

The final version of the main landing gear assembly, which incorporated a landing light. (Lennart Lundh)

The forward-retracting nose wheel landing gear used either a spoked pattern or this slotted version, both of which were used interchangeably. (Author)

bonding metal with adhesive tape versus liquid. The engine occupied the rear portion of the fuselage and was housed in a stainless steel shroud, with titanium used for most of the mounting structure. Externally, the lateral, bulged fairings (which increased the aircraft's total cross-section aft of the wing and kept it constant) and surrounding exhaust area were also of titanium.

The Delta Dagger's huge vertical tail was swept back 52 degrees and incorporated a rudder with an area just under 10 square feet. Due to the thinness of the fin, rudder control

The cavernous forward electronics compartment is seen here being gutted in preparation for conversion to a drone. The open door behind the canopy indicates the location of the upper electronics compartment. (Terry Love)

The F-102's control column incorporated dual grips. The right grip controlled the aircraft, while the left worked independently to operate the radar suite. (Norm Taylor)

linkage ran externally and was covered by a fairing on the fin's port side. The original fin was 104 inches high with an area of 68 square feet, but was increased to 137 inches and 95 square feet. Although the larger fin slightly reduced performance (by Mach 0.1), it solved the problem of roll-coupling. Also incorporated into the tail were two pitot pressure tubes in the leading edge, a UHF antenna at its very tip and a VHF antenna in the form of lateral extensions on each side of the fin.

Located at the base of the fin's trailing edge were speed brakes, commonly called "speed boards." Standard brakes were enlarged 40 percent over earlier types and incorporated a drag chute, which reduced landing speed by more than 30 mph.

The leading edge camber of the earlier Case X wing formed into an upturned wing tip. (Author)

The blending of the canopy into the spine fairing to the base of the fin eliminated airflow disturbances and subsequent turbulence associated with conventional canopies. The canopy itself was lighter and used less framing than the original type. Its swept, knife-edged windshield conformed to the Dagger's rakish lines. A thin, black-painted partition, called a

The TF-102A's engine intakes were faired into fuselage recesses to conform to the area rule. Plainly visible is the intricate pattern of vortex generators on the canopy frame to improve airflow over the broad area and eliminate buffeting. (Author)

The Case XX wing featured a noticeable camber, which continued to the wing tip. Visible atop the forward fuselage spine, from front to rear, are a UHF antenna, retractable red beacon, and IFF antenna. (Author)

During taxi tests and barrier engagement trials at San Diego's Lindbergh Field, the second YF-102A prepares to engage the barrier, which would cause the cable to snap upward to be caught by the main landing gear. The aircraft has a lengthened nose, but still has the original heavy-framed canopy. Also of interest is the natural metal tail fin and one-piece landing gear doors. (General Dynamics)

"vision splitter," was often added to the center windshield frame to prevent bothersome reflections between the flat panes.

The cockpit itself was a standard arrangement which underwent little change throughout developmental phases. Despite the 15-degree downward slope of the nose, forward visibility was limited primarily by the dominant radar scope at the top of the instrument panel. Visibility became more restricted when an optical sight was added above, in what little space remained in the windshield cavity. A single control stick had two grips, the right one of which was used to fly the aircraft, while the offset left grip moved independently to control the radar. Its movement swept range parameters to facilitate target lock-on. The radar scope, which also displayed aircraft attitude, was difficult to view in daylight, forcing the pilot to press his face against a rubber hood, while the throttle was set to military power.

The heart of the Delta Dagger was the Hughes MG-10 fire-control system, which replaced the MG-3 and comprised the AN/ARR-44 data link, the MG-1 automatic flight control

The barrier probe, which was designed to catch the barrier cable, was mounted on the fuselage center-line just forward of the main landing gear. It was lowered simultaneously with deployment of the drag chute. (Author)

The second-built YF-102A is positioned to show how it would be restrained by Convair's safety net at Lindbergh Field. The barrier was available for use during high-speed taxi tests in early 1955. When the nose gear hit the net, a steel cable snapped up to catch the main landing gear. The cable then dragged 30 tons of anchor chain. Convair claimed that the chain would stop any runaway F-102 in less than 600 feet. (General Dynamics)

Proof that the barrier was not the solution to all landing mishaps lies in this view of a 317th FIS F-102A at Elmendorf AFB, Alaska. Following nose gear failure, the pointed nose of the Deuce caught the barrier cable, which snagged the IR sensor. (USAF)

system, and the AN/ARC-34 miniaturized communication set. Interception tactics were quite flexible and usually dependent on particular circumstances. A standard method had a pair of F-102s flying in trail with their fire-control radar searching for targets. After detecting a target, the MG-10 locked on to it and tracked it automatically. With the microwave scanner feeding into the flight control circuits, the computer worked out a collision-course flight path. Meanwhile, the pilot selected either radar or infrared missiles and armed the firing sequence. Three missiles were then automatically readied for firing, and at the precise moment, the missile bay doors opened and the missiles launched. Beginning in 1963, a hypersensitive IR sensor, visible as the distinctive ball atop the nose, was added to greatly enhance the basic IR capability. Data link was added in 1965, which allowed the aircraft to be flown by remote control from the ground. Course and altitude directions could then be fed directly into its autopilot.

The fire-control system accounted for nearly 1,700 pounds of the F-102's weight, and portions of its sighting system had to be cooled with liquid nitrogen. The radar antenna was housed in the distinctive needle nose, while the majority of

The ram air turbine of this 482nd FIS Deuce is visible just forward of the right main landing gear well in the fuselage. (Paul Wagner)

An early version of the F-102 while assigned to the Air Proving Ground Center. (W.J. Balogh, Sr. via David Menard)

Position lights can be seen on the fuselage fairing, just aft of the eleven and on the upper portion of the missile bay door of this 48th FIS F-102 in September 1959. (Merle Olmsted via David McLaren)

A Deuce of the Hawaii Air National Guard deploys its drag chute upon touchdown. (Nick Williams)

Vertical lines in the windshield are electrical heating elements. The antenna, just forward of the nose landing gear door corresponded to a TACAN installation, which was introduced to the F-102 during late 1970. Seen in the cockpit are the vision splitter, radar scope, and optical sight. (L.B. Sides Collection)

fire-control equipment occupied the forward electronics bay between the radome and cockpit. Other electronics bays were located immediately aft of the cockpit, and above and aft of the missile bay.

Other primary features incorporated into the fuselage included a ram-air turbine, which was attached to a door immediately forward of the right landing gear well. The turbine's variable-pitch fan drove a hydraulic pump capable of operating control surfaces, the speed brakes, and landing gear in the event of a main power loss.

Other safety concerns were addressed during the late 1950s when Convair conducted tests to broaden the F-102's safety margin during landings. In the wake of problems encountered with drag chutes, nose gear, and speed brakes, a YF-102A was tested against barriers installed at both ends of runways. When the aircraft's nose landing gear engaged the safety net, a steel cable snapped up to catch the main wheels. Attached to the cable were 30 tons of ship anchor chains, which, according to Convair, was sure to stop any runaway F-102 in less than 600 feet. It was discovered that the dis-

This view of a late model Deuce shows the landing gear well configuration. The dark square immediately aft of the nose landing gear door is the intermediate electronics compartment. (Nick Williams)

F-102A of the 482nd FIS at Seymour-Johnson AFB, North Carolina. (Paul Wagner)

The spoke-patterned nose wheel of this factory-fresh F-102 was one of two types used. (General Dynamics)

tance between the main and nose gear was so great that the cable fell back to the ground before it could catch the main gear. The solution was the addition of a barrier probe under the fuselage, which extended when the drag chute handle was pulled.

Since the wide variety of emergencies dictated that more options be available, a tail hook was added during 1960. The hook was designed to snag a cable of the BAK-9 barrier system, which reeled out a tension tape as braking resistance. The first arrested landing by a USAF aircraft was made by an F-102 of the 525th FIS at Bitburg AB, Germany, on 23 July 1963. In that incident, Captain Leslie J. Prichard's F-102 blew a main tire on a two-ship takeoff, causing him to nearly collide with the lead aircraft. When the decision was made to

use the tail hook to engage the BAK-9, Captain Prichard touched down 1,100 feet from the cable. Severe vibration from the blown tire made nose wheel steering difficult and resulted in the loss of wheel brakes and drag chute just prior to snagging the cable. The hook caught the cable, bringing the aircraft smoothly to a halt in 720 feet. Although not a solution to all landing emergencies, use of the tail hook became common practice.

Power for the F-102 was supplied by a Pratt & Whitney J57-P-23 turbojet with a flap-type nozzle, tandem cylinder, and forged flap afterburner. It was rated at 16,000 pounds static thrust with afterburner and 9,800 pounds in military power. A J-57-P-25, rated at 17,200 pounds with afterburner, was also available. Maximum speed initial climb rate was 825

Following rollout, the first TF-102A was positioned for this comparison view. Note the elevon span of both variants. The small dark circles in the wing tips are navigation lights, while the lateral extension near the top of the tail fin is a UHF AN/ARC-34 antenna. (General Dynamics)

Cockpit of the F-102 flight simulator. (USAF)

The MG-3 mockup of the F-102's fire control system. (USAF)

mph (Mach 1.25) at 40,000 feet and its rate of climb was 13,000 feet per minute. The TF-102A had a maximum speed of 646 mph (Mach 0.97) at 38,000 feet. The F-102A had a combat ceiling of 51,800 feet and a service ceiling of 54,000 feet, while the TF-102A's was 50,000 feet. The Delta Dagger was a hefty fighter, weighing 19,350 pounds empty and 24,500 pounds fully combat ready. Its maximum takeoff weight was 31,500 pounds. A pressurized internal fuel load of 1,085 U.S. gallons allowed a mission range of 1,350 miles, which was extended by the later addition of two 215-gallon underwing tanks. The F-102A's overall length was 68 feet, 3 inches, while the TF-102A was shorter at 63 feet, 4 inches.

From a performance standpoint, the F-102 seemed eager to fly, despite its unseemly girth. During long takeoff rolls,

Details of the forward section of an F-102A of the Pennsylvania Air National Guard. Air Force policy required that a red and white armament placard be applied to the forward fuselage and list all munitions aboard. The landing gear door was painted red, white, and blue. (L.B. Sides Collection)

F-102s undergo depot-level maintenance at the San Antonio Air Material Area at Kelly AFB, Texas, in May 1959. A decade later, most IRAN work was turned over to civilian aerospace firms. (Author)

especially with earlier "straight leg" variants, the runway showed no sign of falling away until the afterburner was lit. Pilots then experienced an awe-inspiring boost best described as being shot out of a cannon. At extreme altitudes, the heavy airplane became very sensitive to control inputs and, thanks to the automatic flight control system, gave little indication of slipping through the sound barrier. The transition from sub-sonic to supersonic flight is not instantaneous, as the term "sound barrier" implies, but occurs through a range of speeds from 600 mph to 900 mph. In mock attacks against SAC B-52 bombers, the F-102 proved itself nimble enough to easily turn inside the agile Stratofortress. Speed and lift during landing could be quickly dissipated by assuming a nose-high approach, which allowed the delta wing to act as an aerodynamic brake.

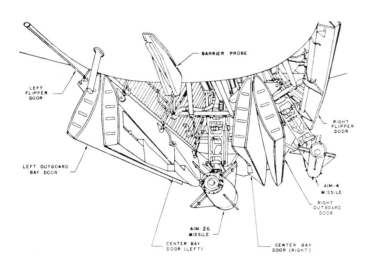

F-102 Missile Bay Modification to accommodate AIM-26 (looking forward)

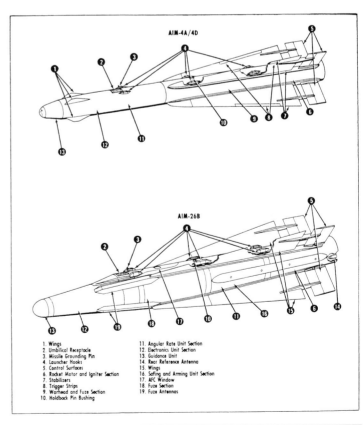

1. Wings
2. Umbilical Receptacle
3. Missile Grounding Pin
4. Launcher Hooks
5. Control Surfaces
6. Rocket Motor and Igniter Section
7. Stabilizers
8. Trigger Strips
9. Warhead and Fuze Section
10. Holdback Pin Bushing
11. Angular Rate Unit Section
12. Electronics Unit Section
13. Guidance Unit
14. Rear Reference Antenna
15. Wings
16. Safing and Arming Unit Section
17. AFC Window
18. Fuze Section
19. Fuze Antennas

An F-102 pilot, dressed for high altitude flying during the early 1950s, poses with the first Falcon missile, designated the YGAR-1. On later models, trailing edge control surfaces were separated from the stabilizers for improved high altitude performance. The rocket's fins folded within the diameter of the tubular body to facilitate loading in launch tubes. (General Dynamics)

ARMAMENT

Although the Air Force quest for the 1954 Interceptor officially began during early 1949, the air-to-air missile system slated for inclusion under the Weapon System Concept had its start two years earlier. In a 1947 proposal, in response to an Air Force search for a new type of armament system, the

The Hughes Falcon family. From left to right: GAR-11, GAR-1D, GAR-2A, AIM-4G, and AIM-4F. (General Dynamics)

Hughes Aircraft Corporation began work on the airborne fire control system and a guided air-to-air missile for interceptor aircraft.

The original MG-3 Radar Fire Control System for the F-102, when updated with automatic flight control and data link subsystems, became the MG-10. The training version installed

In a fiery blast, the third-built F-102A unleashes a salvo of 2.75-inch rockets during a test on 2 October 1957 at Holloman AFB, home of the U.S. Air Force Armament Development Center. Despite the flame impingement on the aircraft's fuselage, a special paint application prevented damage to the F-102's skin. (General Dynamics)

Test pilot James Eastham just prior to an F-102 test flight following modifications to accept the AIM-26A/B Falcon missiles. Alterations to the center missile bay door are evident in comparison to the original door it abuts. (Marty Isham Collection)

Under the watchful eye of judges during a loading competition, a full load of 24 2.75-inch rockets is loaded aboard an F-102A. Cases specially designed for Falcon missiles are in the foreground. (USAF)

An AIM-26 is loaded into the center missile bay. The absence of rocket tubes in the center bay doors indicates the modification made to accommodate the large missile. The smaller upper door was called the "flipper door." The AIM-26 is flanked by a pair of AIM-4Cs. (WI ANG)

in a small number of TF-102As was designated MG-10T. The MG-10's three main functions were to help the pilot locate long-range targets, steer the attack course, and prepare and fire ordnance with precise timing. Data link and the AFCS provided the added functions of flying the aircraft during the attack, automatic flight control during instrument approaches, and autopilot. That left the pilot with the responsibility for take-offs and landings, armament and target selection, maintaining the intercept course, and systems monitoring.

The MG-10 gave the pilot three options for attack: radar-lead collision, in which a missile or rocket attack could be automatically or manually carried out, and optical-lead and optical-pure pursuit, both of which required manual operation of missile or rocket attacks.

The weapon system requirement of the 1954 Interceptor originally specified by the Air Force called for air-to-air missiles, backed up by 2-inch or 2.75-inch rockets, or the T-110 rocket gun. Designed to fire T-131 2.75-inch rockets, the gun was then under development by the Navy and drew Air Force interest when tests revealed a 1,260 feet-per-second muzzle velocity, which guaranteed accuracy.

Like a number of first-generation missiles, the Hughes MX-904 design was given a standard aircraft designation. Originally designated the F-98 "Falcon" at the onset of its joint Hughes-Air Force development, the missile had its origins in the "Tiamat III," an awkward-looking 625-pound radar-guided air-to-air missile. By 1949, Tiamat had slimmed down to a 6-1/2-foot, 120-pound tubular Model A projectile,

The Hughes AIM-4D-8 loaded in an F-102's right missile bay. (Lennart Lundh)

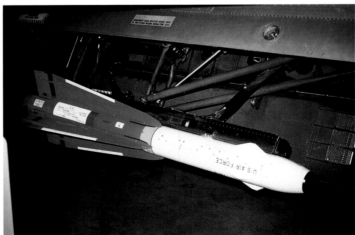

Falcon missiles were connected to a rail launcher attached to a scissors-type apparatus which rapidly extended on command from the fire control system. (Lennart Lundh)

Test pilot Eastham receives last minute instructions prior to a test flight at Holloman AFB during May 1960. A camera pod was added beneath the left fuselage, while an ECM pod for the Falcon tests was carried beneath the right wing. Mission marks of the Hughes tests were recorded on the aircraft's nose. (James Eastham via Marty Isham)

which was a direct forerunner of the GAR-1, first of the Falcon series. Hughes' original Falcon would ultimately yield numerous missile variants, which employed either semi-active radar or infrared (IR) guidance systems. A radar-guided Falcon homed on the return signal bounced off a target by the launching aircraft's radar. An IR Falcon was attracted to the target's radiated heat energy (usually the engine exhaust of enemy aircraft) and was especially effective at distinguishing low altitude targets. Especially appealing was the IR missile's "fire and forget" capability, which meant the launching aircraft could break off immediately after firing. Since both types could be countered by enemy bombers, F-102s carried a variety of missile combinations. It was standard practice to fire both types to ensure the kill.

Built within the F-102's four large missile bay doors were 12 tubes for launching 24 rockets in tandem. (Lennart Lundh)

The Falcon family of missiles was designated GAR (Guided Aerial Rocket) during 1950 and became AIM (Air intercept Missile) under the 1962 redesignation program. As a "direct hit" weapon, the missile was designed to penetrate the target before detonation, ensuring total destruction. The Falcon was powered by a Thiokol solid-propellant, single-stage rocket engine, giving it speeds well beyond Mach 2. Its basic airframe was a tubular magnesium structure, which contained the guidance and control system, electrical and hydraulic systems, rocket motor, and warhead and fuse system. Four large stabilizers with trailing edge control surfaces were Falcon earmarks. The control surfaces were later separated from the wing, which increased maneuverability at high altitudes by 40 percent. Falcons carried by the F-102 averaged a 5-mile range with a 50,000-foot ceiling, depending on the altitude from which they were fired.

Most Falcon tests were conducted at Holloman AFB, New Mexico, where the first ground launching took place in March 1949. North American B-25 bombers served as the primary launch aircraft and radar test platform. The first air launch of a Falcon took place during 1951, and by fall 1952, more than 400 had been expended during extensive evaluations. Northrop's F-89H Scorpion became the first operational interceptor to be equipped with Falcons, with the first launch from an F-89 having occurred in September 1952. It wasn't until early 1955 that the Air Force seemed totally convinced that the F-102 was fully capable of launching Falcon missiles. The first firing from an F-102 took place in May 1955. Armament tests peaked during early July, when a YF-102A launched its full load of six Falcons and 24 rockets in less than ten seconds.

During the mid-1950s, the Air Force considered the feasibility of giving the F-102 a nuclear weapon capability. The Douglas MB-1 "Genie," then undergoing trials with the F-89 Scorpion, was less complex than other missiles since it did not require a guidance system. As such, it was technically an unguided rocket with a blast intensity capable of destroying a group of bombers with merely a near miss. The first launch of a Genie from an F-102 took place during May 1956. However, since its inclusion in the Delta Dagger's weapons package would mean production delays, the Air Force gave up on the idea by early 1957.

Meanwhile, Hughes proceeded with a nuclear version of the Falcon, designed specifically for the F-102. First known as the GAR-11 and later redesignated the AIM-26A, the nuclear Falcon was introduced during 1960. Its atomic warhead was similar to that of the Genie, with triggering accomplished by active radar proximity fuse aerials located forward of the stabilizers. Since the GAR-11 was somewhat larger than the standard Falcon, the six rocket tubes were removed from the F-102's center bay doors to accommodate the oversize missile. Other changes included strengthening of the missile bay roof structure and beefing up the center launcher mechanism to support the GAR-11's added weight. By 1963, nearly half of the F-102s produced had been modified to carry the nuclear Falcon. Later modifications gave the Deuce more armament flexibility by allowing a mix of AIM-26As and AIM-4s to be carried in the center missile bay.

Both modified and unmodified F-102s featured six missile assemblies: three abreast in the forward section of the missile bay and three in the rearward portion. Load configurations depended upon modifications, with a wide variety of combinations possible. The firing sequence was extremely fast-acting. On command from the fire control system, the missile bay doors snapped open, the scissors-type assemblies extended simultaneously, and the three missiles were automatically fired at 50-millisecond intervals. Immediately after launch, the assembly retracted and the other group of three Falcons extended to the firing position. If rockets were selected, only the doors opened. Covering the missile bay were six air-operated doors, each of which housed six rockets placed tandem in three tubes, and two smaller outer doors, called "flipper doors."

The Air Force decided during August 1950 that the F-102's weapon system should include rockets as backup for the missiles. The original arrangement of 36 2-inch T214 rockets gave way to 24 2.75-inch Folding Fin Aerial Rockets (FFAR). The unguided rocket was 48 inches long and weighed 19 pounds with a high-explosive warhead. Although a safety switch was installed to prevent accidental firing of the aft rocket, often only the forward rocket was loaded, much to the crew's relief. Shortly after the rocket capability was reduced to half to make room for the AIM-26 missile, their use as secondary armament was discontinued.

Falcon production began in 1954 with more than 50,000 missiles built at Hughes' Tucson, Arizona, facility. Production ended in 1963, however, modifications continued until 1969, and many were periodically upgraded. Several thousand additional Falcons were produced in Sweden under a licensing agreement with SAAB-Scania. Ten major variants were operational with the F-89, F-101, F-102, F-106, and F-4 aircraft. The AIM-4 family of Falcons used by the F-102 were also standard armament for the F-89H and F-101B. A Super Falcon series was developed to fulfill mission requirements of the F-106, and AIM-4Ds went to war in Southeast Asia aboard F-4 Phantoms under the "Dancing Falcon" program. Their disappointing performance in dogfights led to the enhanced, but short-lived, AIM-4H.

More than three decades after the Falcon became operational, members of its family equipped the arsenals of the U.S. and several allied nations. The Falcon's profile can still be seen in the AIM-54 "Phoenix" and AGM-65 "Maverick" missiles. Falcon missiles that were operational with the F-102 are described as follows:

GAR-1/AIM-4 - The forerunner of the Falcon family, it was originally intended to arm the F-89 on an interim basis until the arrival of the F-102B Ultimate Interceptor (F-106). The radar-guided missile could attain speeds exceeding Mach 2. It was 78 inches long, 6.4 inches in diameter, had a stabilizer span of 20 inches and a prelaunch weight of 127 pounds. It entered service with the F-89H. More than 4,000 were produced from 1955 to 1961.

GAR-1B - Basically a GAR-1 converted with an IR seeker head and associated circuitry.

Armament specialists of the 57th FIS load Falcon missiles. A yellow high explosive band was applied to the center of the missile body, while the brown band farther aft signified low explosives. (USAF)

GAR-1D/AIM-4A - Although its dimensions were identical to the GAR-I, it featured control surfaces which were noticeably separated from the stabilizers, which became the standard for all subsequent models. The radar-guided missile entered service with the F-89H, F-101B, and F-102A in 1956. A thorough operational evaluation of the GAR-1D in 1957, labeled Operation FAST DRAW, led to a sizable production count of more than 12,000 by 1961.

GAR-2/AIM-4B - An IR-guided version of the GAR-1. The first example was launched in April 1954 and became operational with the F-89H, F-101B, and F-102A in 1956. Although considered an interim model, more than 1,600 were produced.

GAR-2A/AIM-4C - Similar to the GAR-2, but slightly heavier as a result of minor improvements It entered service with the F-101B and F-102A and later with the F-106. A total of 9,500 were produced from 1956 to 1960.

GAR-2B/AIM-4D - An upgraded IR-guided combination of the advanced AIM-4G (familiar to the F-106) IR seeker and the AIM-4C airframe. Introduced in 1962, most were converted from earlier GAR-1Ds and -2As. It weighed 134 pounds and had a top speed of Mach 4. It was manufactured in Sweden as the Rb 28 for the J-35 Draken and AJ-37 Viggen. Its international appeal also led to its use on Royal Canadian Air Force F-101s and Japanese Air Self Defense Force F-4Js. The type went on to equip F-4s during the war in Southeast Asia, but was quickly replaced with the AIM-9 "Sidewinder" after a dismal showing in air-to-air combat.

GAR-11/AIM-26A - A radar-guided Falcon with a nuclear warhead similar to that of the earlier MB-1 Genie. Considerably larger than the AIM-4 series, the AIM-26A had a length of 84 inches, a diameter of 11 inches, stabilizers that spanned 23-1/2 inches, and a weight of 203 pounds.

GAR-11B/AIM-26B - Basically an AIM-26A with a conventional high-explosive warhead. Although shorter than its atomic predecessor, its diameter was increased to 11-1/2 inches and it was heavier at 262 pounds. Both versions equipped the F-101B and F-102A, and later, Royal Swedish Air Force J-35F Drakens and Mirage 111.S fighters of the Swiss Air Force. It was differentiated from the -26A low-explosives by a yellow high-explosives band around the warhead and a brown band around the rocket motor station. Early models displayed an orange band on the forward portion of the missile body.

COLOR SCHEMES AND MARKINGS

The F-102's rakish lines and broad surfaces lent themselves to the application of interesting and colorful markings. Those seen on aircraft that filled the ranks of front-line units reflected not only true fighter squadron spirit, but a period in U.S. Air Force history during which such practice flourished. The Delta Dagger was, after all, a stylish airplane and looked good in whatever it wore—a point quickly argued by many in view of the introduction of camouflage.

Nearly all F-102s came from the factory painted overall glossy Aircraft Gray (FS 16473), which was often called "Air Force Gray", but best known as "ADC Gray." The neutral tone became the Air Force standard, which not only presented a uniform finish, but protected aircraft surfaces. Its application to the F-102 was vital in shielding the interceptor's skin from the rocket blast of Falcon missiles. Residue form the missile's rocket motor was inert until it got wet, then moisture triggered an electrolytic corrosive action that broke down metal alloys. To ensure adequate protection, the F-102 was given a coat of wash primer and a coat of zinc chromate primer, followed by a coat of the high-grade enamel. Magnesium components received an extra undercoat of zinc chromate, while titanium parts remained unpainted.

The only exceptions to the ADC Gray livery were early variants committed to test programs and 50 F-102s involved

Common among the wide variety of special markings worn by many F-102s committed to tests were large red-painted areas. They were returned to standard Air Force gray finish during the Test-to-Flight period. This Deuce was used for tests by the Air Research and Development Center at Wright-Patterson AFB, Ohio. (Larry Davis Collection)

"Day-glo" stripes were applied to Deuces involved with "Operation Southern Tip," an Air Defense Command exercise. These banded F-102As of the 325th FIS, 327th FG are seen at McEntire ANGB, South Carolina, during August 1963. (Marty Isham Collection)

In many cases, the markings used to identify a particular unit were continued on the external fuel tanks. The tanks of this California Air National Guard Deuce denote the 144th Fighter Group as the parent unit. (Nick Williams)

in a temporary switch to silver acrylic lacquer. By 1969, depot-level F-102 maintenance (which was performed primarily by civilian aerospace firms) was shifted to Fairchild-Hiller at Bob Sikes Field at Crestview, Florida. An aircraft spent three to five months at the facility before it returned to its home unit. Of 67 F-102s that arrived for overhauls between 12 May and 30 October 1969, 50 received the silver finish on an experimental basis. The Air Force reverted to ADC Gray when it was determined that the silver acrylic lacquer had less protective qualities. Some of the silver-painted F-102s were eventually repainted gray.

This F-102A may have worn the unusual black and gray camouflage during the time that it was an instructional airframe. (Baldur Sveinsson Collection)

Although the first YF-102A featured an improved canopy, the remaining three of the type retained the heavily-framed canopy familiar to the YF-102. Most of the original Deuces wore various patterns of red and white on the nose, which incorporated the "Convair" name. (Interair)

Yellow wing fences and the California Air Guard's flashy tail markings accentuated this F-102A's silver finish. (Marty Isham Collection)

The initial batch of F-102s committed to test programs were usually highlighted with Insignia Red or Fluorescent Red-Orange. The latter also appeared on a small number of operational aircraft, however, its inability to withstand weathering effects outweighed the benefit of high visibility. In keeping with Air Force policy, Deuces that operated over the snowy northern regions featured large red-painted areas for maximum conspicuousness in the event they were forced down. Red-tailed units were the Alaska-based 31st and 317th Fighter Interceptor Squadrons, the 37th in Vermont, the 57th in Iceland, the 59th in Labrador, the 327th (when it deployed to Greenland), and the Wisconsin and Minnesota Air National Guard.

Three variations of standard markings were used, all of which incorporated the required "buzz numbers." Introduced after World War II to identify USAF aircraft, buzz numbers were a combination of a two-letter code identifying the aircraft type (FF for F-102As and TF for TF-102As) and the last

Aircraft sent to the Southeast Asian combat zone during the mid-1960s, such as this F-102A of the 509th FIS, set the standard for U.S. Air force aircraft schemes worldwide. Eventually, most F-102s were finished in the familiar "tri-tone" camouflage. (Terry Love)

When ordered to deploy to Thule Air Base, Greenland, the 327th FIS had large portions of their aircraft painted red, a standard Air Force practice for aircraft that operated in arctic climates. Here, Deuces of the "Iron Mask" squadron are lined up at Ernest Harmon AFB, Newfoundland, during the Thule deployment in June 1958. (Budd Butcher)

As one of the few F-102 units that sported high-visibility red tails, the 317th FIS transferred the majority of its aircraft to the Wisconsin Air Guard when it was inactivated. The 176th FIS retained the markings in view of Wisconsin's propensity for harsh winters and operations along the northern tier. (Hugh Muir via Terry Love)

Flying off the wing of an Air National Guard F-89D, a Deuce of the Minnesota Air Guard wears Fluorescent Red-Orange, which was eventually replaced in favor of Insignia Red. (Terry Love Collection)

three digits of the aircraft serial number. All standard markings were of "Scotchcal" pressure-sensitive film, which was impervious to chemicals, supersonic speeds, and temperature extremes. Air Force painting directives called for flat black anti-glare panels, a gloss black radome and medium gray cockpit interior. The interior surfaces of the speed brakes and missile bay and doors was interior green, while landing gear wells and inner door surfaces were painted zinc chromate.

The application of squadron markings was based on unit policy and ranged from imperceptible insignia to large portions of the aircraft painted to identify a particular unit. Colors and figures were displayed on tail fins, speed brakes, wing fences, drop tanks, landing gear doors, and air intake ramps. Often, the flights within a squadron wore their own distinctive marks. Aircraft flown by group commanders, squadron commanders, and executive officers were routinely adorned with

With its sultry nose art and whitewall tires, this Deuce of the 59th FIS exemplified the F-102's hotrod image. One can only wonder if a pair of fuzzy dice hung in the cockpit. (Marty Isham Collection)

Like this example in storage during 1979, a large number of ANG F-102s extended their service careers as full-scale aerial targets in the drone program, where they were usually repainted. (Terry Love)

red or yellow "commander's stripes," which were applied diagonally to the fuselage. Personal markings and nose art were commonly observed on F-102s. Air National Guard color schemes and markings were as expressive as those worn by their Air Defense Command counterparts.

The darkening situation in Southeast Asia marked the end of the era during which Air Force aircraft proudly displayed their distinctive liveries. Soon after F-102s were deployed to the Asian theater, the ranks of brightly marked F-102s dwindled until nearly all wore camouflage. The standard scheme for all USAF aircraft types operating in Southeast Asia consisted of two shades of flat green (FS 34102 and FS 34079) and flat tan (FS 30219), with flat gray (FS 36622) undersurfaces. Radomes became flat black, and national markings were reduced to a mere 15-inch size. The only other markings worn were white or black "USAF" titles and serial numbers on the tail fin. Application of the "tri-tone" camouflage became part of the IRAN process as aircraft were rotated through maintenance facilities. In addition, it became standard practice for commanders of Air National Guard units

Seen here at McClellan AFB, California, during 1966, the first YF-102A may have been involved with camouflage tests while assigned to the Air Force Operations Group. The radome was left unpainted. (General Dynamics)

This F-102A displays the original fuselage markings. (General Dynamics)

The 179th FIS of the Minnesota Air Guard inherited most of their F-102s from the Labrador-based 59th FIS when it was inactivated. The 179th retained most of the 59th's red markings since it too operated along the northern tier. The 179th deviated from the usual practice of assigning aircraft individual squadron numbers, using letters worn on the speed brakes instead, (Maj. Gen. Wayne C. Gatlin)

to request that their F-102s be camouflaged during scheduled maintenance cycles.

Under Military Assistance Program guidelines, F-102s that were destined for Greece and Turkey were painted ADC Gray during their pre-delivery overhauls. For F-102s relegated to the drone program, painting guidelines became a moot issue, since time and money would only be wasted on aircraft earmarked for destruction.

The Pennsylvania Air National Guard used this striking black pattern for its F-102s. The aircraft number on the tail fin tip of this late model Deuce was repeated on the wing elevon. The external fuel tanks were also black and bore the parent 82nd Fighter Group title and Air Force Outstanding Unit Award. Col. Ed Bollen, commanding officer of the 112th Fighter Group at Pittsburgh and Lt. Grant Bollen, his son, teamed up for this training flight on 19 December 1973. (Marty Isham Collection)

Fuselage markings were later reduced and repositioned farther aft, with the national insignia relocated to the intake housing. This Deuce of the 61st FIS was trimmed with red. The pilot's name was applied to the left side of the canopy frame, while the crew chief's appeared on the right. (Jerry Geer Collection)

Some early variants wore this non-standard marking scheme. (General Dynamics)

Designated a JF-102A during tests, this early Deuce underwent cold weather tests at Ladd AFB during the winter of 1955-56. It is seen here in August 1957 while assigned to the ARDC at Wright-Patterson AFB. (USAF)

The F-102 pioneered the use of tail codes. Not only did this Deuce sport the first tail code for the F-102, but the first for the Air Force. While assigned to the 6520th Test Group at Hanscom AFB, MA, during the early 1960s, this JF-102A was trimmed with white stars on red backgrounds and wore its name, "Texas Terror," below a pair of steer horns on the nose. (T. Cuddy via Marty Isham)

Typical of F-102s flown by unit commanders was this example adorned with red diagonal stripes. The pilot and 141st Fighter Group commander, Col. Charles E. Nelson, Jr., is noted on the nose landing gear door, while the emblem of the 116th FIS appears on the fuselage. (Marty Isham Collection)

Above: Finished in overall "ADC Gray," the standard for F-102s, a trio of factory-fresh Deuces of the 327th FIS flies over California on 6 May 1957. (USAF)

Silver-Painted F-102s
(In order of their arrival at Crestview)

56-1416	56-0998	56-1401	56-1102	56-1109
56-2356	56-1365	56-1132	56-2344	54-1407
56-1466	57-0870	57-0775	56-2336	56-1330
56-1376	53-1803	56-1134	56-1336	56-1331
56-1453	54-1387	57-0806	56-1306	56-1474
56-1123	56-2334	56-2378	56-1471	55-3427
56-1446	56-1003	56-1447	56-1413	57-0792
54-1393	56-1369	56-1418	57-0847	54-1365
56-2340	53-1817	57-0786	56-0990	56-2351
53-1802	56-1017	57-0883	56-1443	57-0828

Taken more than eight years apart, these two views illustrate the dramatic difference between how number 980 looked in 1965, while assigned to the 32nd FIS, and in its twilight with the Florida Air Guard. (Norm Taylor)

Air Defense

The air defense of North America seems to have had its humble beginnings during the final months of World War I, when great strides were made in aerial bombardment. Although airpower proponent General William Mitchell foretold of the possibility of bomber attack from Europe, America seemed secure in the fact that its ocean expanses offered sufficient protection from such danger. With the rapid advancement of the airplane and the rise of Japan and Nazi Germany during the 1930s came the realization that North America was indeed vulnerable to air attack. When General Headquarters Air Force was created in 1935, its mission statement included the requirement to "Provide the necessary close-in air defense of the most vulnerable and important points in the United States."

When war erupted in Europe in September 1939, the U.S. War Department did little to bolster America's meager air defenses. Military leaders were convinced that the U.S. was safe from bomber attack, which they felt was best prevented by not allowing adversaries to establish bases in the Western Hemisphere. Nevertheless, the War Department established an Air Defense Command in February 1940. Located at Mitchell Field, New York, it comprised elements from the Air Corps, Coast Artillery, and Signal Corps.

During early 1941, air defense was turned over to the air forces, which divided it into four continental air forces, each with an interceptor command. Its work essentially completed, the ADC was inactivated during June. At the outbreak of World War II, the elements necessary for an effective air defense—pursuit plane squadrons, an early warning system, and radar—were in place. Although the ADC did not officially exist, its doctrines guided air defense operations throughout the war.

Since U.S. air defenses virtually ceased to exist at the end of World War II, the Air Defense Command was necessarily established as one of three combat commands in March 1946. Although air defense was perceived as a vital compo-

Since the Air Defense Command suffered a severe aircraft shortage as a result of the Korean war, a plan was devised to augment the ADC's runway alert program with Air National Guard aircraft. The 194th Fighter Squadron of the California Air Guard, which received F-51Hs during mid-1952, pioneered the program in 1953. The following year, the unit's Mustangs were replaced by F-86A Sabres. (Gordon S. Williams)

World War II vintage Republic F-47s, along with F-51s, flew alongside the ADC's early jets until replaced by the F-86 Sabre. (Brian Baker)

The Lockheed F-94 Starfire was a first generation interceptor, which entered air defense service in May 1950. Plagued by severe maintenance problems, the majority of F-94s remained grounded for extended periods, which, in turn, weakened the air defense system. (Gordon S. Williams)

The F-86D Sabre interceptor, which was replaced by the F-102, had a top speed of Mach .9. It was fully operational during mid-1953, the time frame originally slated for completion of the F-102. (Brian Baker)

Displaying all of the original characteristics of the original F-102 version, this Deuce entered service with the Air Defense Command's 86th Fighter Interceptor Squadron in 1957. (General Dynamics)

An F-102A of the 325th FIS displays its load of Falcon missiles at Truax Field during May 1963. The 325th was subordinate to the 325th Fighter Group of the 30th Air Division. Centered in the unit's red and white tail markings is the ADC emblem. (Leo Kohn)

The 327th FIS undergoes an inspection at Thule AB in September 1958. The squadron deployed with 12 F-102As and 2 TF-102As during June of that year, becoming the first Deuce unit to serve abroad. (Budd Butcher)

One of the first F-102As delivered to the 327th FIS, this early model Deuce wears the unit's "Iron Mask" emblem on the nose. Before long, the 327th changed to more distinctive markings and would adopt three additional marking schemes before it was inactivated in 1960. (Larry Davis Collection)

Wearing stripes reserved for commanding officers' aircraft, this 327th "short-tail" Deuce also sported a meticulously-applied checked pattern. This F-102A is seen at the Oklahoma City airshow in 1956 during a period in Air Force history when such distinctive markings flourished. (Budd Butcher)

North American F-86D Sabres dot the background at George AFB as F-102As of the 327th are prepared for the day's flying. The Deuce in the foreground wears the second marking scheme used by the unit. With the squadron divided into four flights, each wore the symbol of a card suit on the speed brakes. The Deuce in the background had yet to be modified with intake ramps. (Budd Butcher)

Despite busy flying and training schedules, air defense fighter interceptor crews found time to relax. Here, pilots (and friend) of the 327th FIS at Thule AB, Greenland, gather at their makeshift club, named after the town near their home base, George AFB, California. (Budd Butcher)

nent of the postwar air force, severe budget constraints prevented its acquisition of new equipment. Funding that was available was typically channeled to the Strategic Air Command and the Tactical Air Command. It took the precarious nature of Cold War events in 1948 to alert the government to severe shortcomings in the ADC. A partial solution came from pooling the ADC's scant aircraft inventory with that of the Tactical Air Command. Both TAC and ADC were assigned to the newly activated Continental Air Command during December 1948.

Air defense was viewed in a whole new light when Russia detonated its first atomic bomb during September 1949. The event changed forever the American ideology concerning its world dominance. Nevertheless, the on-again off-again status of the ADC continued. The command was inactivated in July 1950, however, when the Korean war revealed the potential to spread globally, it was returned to major command status on 1 January 1951 and relocated from Mitchell AFB to Ent AFB, Colorado. The war's outbreak also caused the acceleration of construction of a permanent radar system, which had begun in 1949. By 1954, the joint U.S.-Canadian Pinetree Line provided extended radar coverage, and the electronic fence that spanned the top of the continent, known as the Distant Early Warning (DEW) Line, became operational in 1958.

To govern the air defense components of the Air Force, Army, and Navy, the Continental Air Defense Command (CONAD) was established during September 1954. Unsurprisingly, the dawning of the jet age changed the face of air defense operations. Rudimentary jets, such as the F-80 and F-84, were replaced in favor of improved types, such as the F-86, F-89, and F-94. The Continental Air Command

Wearing high-visibility markings and equipped with external fuel tanks in preparation for their deployment to Thule, 327th Deuces line up prior to departure. (Budd Butcher)

Deuces of the 327th stop at Frobisher Bay in Canada's Northwest Territory, one of the many stops along the deployment route to Thule, Greenland, in June 1958. (Budd Butcher)

Convair F-102s replaced F-89J Scorpions of the 59th FIS based at Goose Air Base, Labrador in 1960. In addition to the unit's blue trim color, 59th aircraft wore high-visibility markings in keeping with Air Force policy for units that operated in snowy regions. This TF-102A of the 59th FIS is seen at Andrews AFB during June 1965. (Richard Sullivan via Stephen Miller)

This view of a trio of 327th FIS F-102As over Greenland's forbidding landscape shows how the red markings were continued under the fuselage. (Budd Butcher)

had taken delivery of the F-89 and F-94 in June and July of 1950, however, neither proved fully capable of contending with the Soviet bomber threat. When modified to accept the Falcon missile, the F-89D became the F-89H, and F-89J when armed with the nuclear Genie missile. The F-86D was upgraded to F-86L, which was introduced with the F-89H in 1956. The F-89J appeared the following year. Even more advanced was the Century Series supersonic fighter/interceptors. First came the F-100, followed by the F-102, F-104, F-101, F-105, and F-106. As parallel strides were made in radar and weapons technology, it became necessary to assist the pilot with an electronic command and control network that gave warning of an incoming threat and automatically directed the in-

terceptor to the target. Known as "SAGE," for "SemiAutomatic Ground Environment," the first facility became operational during July 1958 at McGuire AFB, New Jersey. More precise ground control took the form of data link, first used with the F-86L.

During the early 1950s, U.S. Air Force and Royal Canadian Air Force leaders addressed the issue of continental defense. The talks culminated with an agreement of joint operational control of air defense forces under the newly formed North American Air Defense Command (NORAD). The command was activated at Ent AFB on 12 September 1957, and by the end of the decade, was fast becoming a vast, multifaceted, interwoven network of sensor sites, control centers, interceptor units, and surface-to-air missiles.

Wearing clothing typical for Thule AB, 327th crews pose with one of their F-102As in March 1959. (Budd Butcher)

Pilots of the 327th FIS test survival suits in the frigid waters near Thule Air Base during the summer of 1958. (Budd Butcher)

Maj Butcher of the 327th FIS wearing a Mark IV survival suit during the unit's deployment to Greenland. (Budd Butcher)

Finished in a striking paint scheme, a 37th FIS F-102 flies over Tyndall AFB in 1959. (R. Goertz via Marty Isham)

The character of the Soviet threat began to change during the 1960s as Russia placed more emphasis on the development of intercontinental and submarine-launched ballistic missiles. The U.S. countered with additional coastal radar and a network of Ballistic Missile Early Warning Sites (BMEWS). When the U.S. and Soviet Union entered the space race, the potential for military operations from beyond the earth's atmosphere was not lost on either. In keeping with the expanding space mission, the Air Defense Command became the Aerospace Defense Command on 15 January 1968. The emphasis on ballistic missile detection and space surveillance became paramount to the extent that the air defense mission went into decline. Citing economic woes, the Department of Defense made drastic cuts in long-range radar sites and the number of interceptor squadrons, beginning in 1963. The DoD's modernization plans, however, were curtailed by technical development problems and the low priority given air defense as a result of the nation's preoccupation with the war in Southeast Asia. The Aerospace Defense Command survived as a separate entity until October 1979, when its interceptor aircraft were transferred to the Tactical Air Command.

Deuce pilots often closed the distance with Russian aircraft to the extent that they felt the vibration from the bomber's contra-rotating props. The 327th FIS intercepted and photographed this "Bear D" near Iceland. (Budd Butcher)

With a view that money couldn't buy, 327th FIS pilots guide their Deuces over glacial Arctic regions. (Budd Butcher)

One of many F-102s assigned to the Air Defense Weapons Center at Tyndall AFB, Florida. The ADWC was established during October 1967 and became responsible for the F-102 weapons training syllabus. (Nick Williams)

A TF-102A of the ADWC at Andrews AFB in December 1968. (Frank MacSorley via Stephen Miller)

To facilitate operational control, U.S. air defense was divided into three zones. On 1 September 1949, the Eastern Air Defense Force was established, followed by the Western Air Defense Force on 21 October. A Central Air Defense Force was activated on 1 March 1951. Within those regions, ten air divisions were established to manage specific areas of responsibility. By October 1955, a total of 17 air divisions were in place. Under a reorganization of the ADC in February 1953, newly created air defense wings were assigned to air divisions. Beginning in August 1956, all ADC wings that operated a control center were designated Air Defense Sectors, which were identified by geographical titles. The year 1956 also saw the Alaskan Air Command and the Northeastern Air Command integrated into the CONAD. At the beginning of 1959, a number of realignments took place as SAGE systems replaced manual control stations. With budget cuts and the shifting of responsibility for air defense to the Air National Guard during the 1960s, air divisions and air defense sectors were systematically discontinued.

When the Air Defense Command was reconstituted during January 1951, it inherited a force comprising 21 fighter squadrons at 14 bases from the Continental Air Command. Two additional air defense squadrons were quickly formed. A total of 21 ANG squadrons were federalized for the air defense mission, giving the ADC control of 44 squadrons at 36 bases by April 1951. Since the ADC suffered a severe shortage of aircraft and crews as a result of the Korean war, a plan was implemented for Air National Guard units to augment the ADC's runway alert program. The aircraft inventory at that time consisted of World War II era fighters, mainly F-51s and F-47s, which flew alongside F-80s and F-84s. Since the Mustangs and Thunderbolts were past their prime and the Shooting Stars and Thunderjets soon reached their maximum potential, they were replaced by F-86 Sabres, F-89 Scorpions, and F-94 Starfires. The success of experimentally assigning ANG units to the ADC's runway alert program led to a permanent assignment under the Air Defense Augmentation Program, which was implemented on 15 August 1954.

Wearing the distinctive markings of the 460th FIS, one of the last Deuces produced lifts off from the Portland, Oregon, Airport. (USAF)

F-102As of the 327th FIS during 1958. (Author)

The "Iron Mask" squadron boasted its mark as the ADC's first supersonic squadron on this sign attached to the Operations building at Thule AB. (Budd Butcher)

With the temperature at 35 degrees below zero Fahrenheit, F-102s of the 327th FIS prepare for a mission from Thule Air Base in 1959. (Budd Butcher)

Additional assistance came from the Navy, whose active concession to the Continental Air Defense Command was a single squadron equipped with new delta-wing Douglas F4D-1 "Skyrays," whose mission was air defense of California's southern sector. The Pacific Fleet All-weather Training Unit (FAWTUPAC) was the counterpart to the Air Force's 27th Air Division at Norton AFB. Operating from NAS North Island, but operationally controlled by the 27th AD, the squadron became the CONAD's first Navy fighter interceptor squadron.

The expectation that the next generation of Soviet bombers would possess the ability to cruise at 500 to 600 mph at altitudes up to 50,000 feet emphasized the glaring need for a very special brand of interceptor. Clearly, the next generation of fighters had to fly in supersonic speed ranges and at extreme altitudes to eliminate every chance of an enemy bomber reaching its target with a nuclear weapon. Of great concern was that America's counterbalance, specifically the Air De-

fense Command, was equipped mainly with aircraft passed from other commands and converted to makeshift interceptors. Particularly disturbing was the absence in the ADC of a single jet-powered, all-weather interceptor specifically designed for its mission. The solution lay in the F-102 among a line of fighters, which came to be known as the "Century Series." The title was a natural spin-off from the designation of the first of the new generation, the F-100 "Super Sabre," built by North American as a day fighter replacement for the F-86. The F-102 was earmarked as an all-weather interceptor, and the McDonnell F-101 "Voodoo" was designed as a long-range escort fighter for the Strategic Air Command. Lockheed's F-104 "Starfighter" was conceived as a high-speed day fighter, while the Republic F-105 "Thunderchief" would replace the F-84 as a fighter-bomber. The F-106 evolved from Convair's F-102.

On 7 January 1953, ADC officials requested that an Advanced Medium-Range Interceptor (MRIX) be developed to

Pilots of the 327th FIS undergo survival training in Greenland during July 1958. (Budd Butcher)

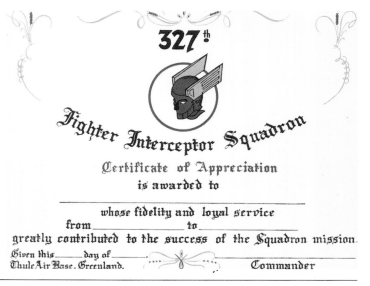

Know all men by these presents that

1st LT. REX D. HOWERTON

has qualified as a

DELTA PILOT

by flying

THE USAF F-102 DELTA WING SUPERSONIC INTERCEPTOR 3 July 1958

FIRST DELTA PILOT

A pilot of the 57th FIS poses with his camouflaged F-102A at Keflavik, Iceland, home base of the "Black Knights" squadron. The visor cover of the pilot's helmet, perched near the cockpit, is checked to match the aircraft's rudder. (Baldur Sveinsson)

replace the F-102 by late 1959. Only 17 F-102s were assigned to the ADC during 1953, all of which were committed to the extensive test program. More were delivered during 1954, and on 28 April, Maj. Gen. Albert Boyd, the WADF commander, became the first operational military pilot to fly the F-102. On 24 April 1956, shortly after flight testing of the F-102's MG-3 radar system, the first tactical F-102A was delivered to the 327th Fighter Interceptor Squadron at George AFB, California. The 1954 Interceptor had finally arrived—two years late. By year's end, a total of 97 F-102s were assigned to five ADC squadrons, a small number compared to 60 squadrons flying primarily F-86s and F-89s. Sixteen additional squadrons converted to the Deuce during 1957, bringing the total number of F-102s in the ADC to 428 by year's end. By that time, intelligence reports indicated that the number of Soviet turbojet-powered Tu-16 "Badger" bombers had increased to 650, Mi-4 "Bisons" to 200, and Tu-20 "Bears"

then numbered 150. U.S. defense analysts calculated that Badgers, which they compared to the Boeing B-47 "Stratojet" bomber, could easily reach targets in Alaska, and with aerial refueling, targets in the northeastern U.S. A total of 2,000 Tu-16s were eventually produced.

The number of F-102s in the Air Defense Command peaked by the end of 1958 with 651 aircraft equipping 27 fighter interceptor squadrons. Such year-end statistics belied the number of F-102s that were actually combat ready, underscoring the interceptor's problematic career in the ADC.

Fielding the F-102, in itself, proved irksome. For the length and intensity of tests involving scores of F-102s prior to operational duty, the Deuce suffered an inordinate number of difficulties. Of great concern were mechanical problems, which plagued the F-102 during its uneasy start with the ADC. Even starting the Deuce held a certain degree of risk. A common source of grief were aircraft starters, which had a tendency to

Underscoring the F-102's high accident rate, this was all that remained of a 20th Air Division F-102A following a takeoff abort on 10 June 1959. (Author)

In this well-publicized photo, an F-102A of the 329th FIS fires half of its Falcon missiles. (USAF)

A Deuce of the 57th FIS flies off the wing of a Soviet "Bear" bomber in 1972. (USAF)

This 57th Deuce sustained damage from a crash on 15 September 1966 to the extent that it was struck from the inventory. (USAF)

overspeed and explode if they failed to engage. The result was often shrapnel damage to the aircraft and those parked nearby, not to mention ground crew, who quickly learned to seek cover during start procedures. F-102s were grounded twice for the problem, which was remedied by encasing the starter to prevent its disintegration.

The Deuce's landing gear also proved to be a common source of trouble, which kept maintenance personnel and Convair technical representatives busy. Hydraulic pistons on landing gear actuators had a high failure rate until it was discovered that wrapping them in fiberglass eliminated the vexation. Nose wheel steering solenoids were prone to freezing, which forced the nose wheel to a full left position, causing distressful moments for the pilot during taxi and landings. Cold weather operations also played havoc with drag chutes, which failed to deploy if water leaking into the parachute compart-

ment froze. In contrast, it was discovered in hot climates that the pressurized air tank in the left wing had a nasty habit of exploding and tearing downward through the wing.

"Screech liners" in the aft engine section frequently burned through the fuselage, resulting in a temporary restriction of afterburner use, except for alert scrambles and test flights. The restriction, in turn, necessitated long takeoff rolls, which meant a higher incidence of tire failure. And at some bases, F-102s were unable to take off during hot weather due to extended takeoff distances.

Other problems in fielding the Deuce presented themselves in a number of ways. The sheer size of the F-102 meant that it didn't fit in existing hangars, called "alert barns." Odd-shaped projections on the front and rear of hangars at ADC bases were obvious signs of where the Deuce made its home. Tie-down bridles to facilitate tests in afterburner mode were

A pair of F-102s of the 57th FIS escorts a U.S. Navy SP-2E "Neptune" patrol bomber over Iceland. Like the Deuce, the Neptune was a key player in the Cold War surveillance game, searching for and tracking Soviet submarines in the world's oceans. (Baldur Sveinsson)

This Soviet "Bear C" bomber was photographed from a 327th FIS F-102 during an intercept mission near Iceland on 4 March 1972. (Budd Butcher)

An F-102A of the 4780th Air Defense Wing at Perrin AFB, Texas. In 1962, the wing was absorbed into the 73rd Air Division, which became responsible for all interceptor pilot qualification. (Jack M. Friell)

By the end of the 1960s, most F-102s remaining in the Air Force inventory had been camouflaged, including those used for training, as evidenced by this lineup of 4780th ADW Deuces at Perrin AFB in August 1970. (Jerry Geer)

in short supply, as were ground support vehicles. Technical manuals for various systems were often unavailable, forcing crews to learn many of the complex systems by trial and error, often a dangerous proposition.

Further complicating matters were the different versions of the F-102, each of which required matching support equipment. Most relegated to the original test program were partially upgraded for tactical use and delivered to front-line units with mismatched components or, in some cases, additional equipment left over from testing. In view of these difficulties, air defense squadrons found themselves hard-pressed to settle into alert and training regimens, a condition further aggravated by F-102s pulled into the IRAN program. To keep the Delta Dagger in a peak state of readiness, a tiered main-

tenance system was used. The immediate level, which covered routine maintenance, involved the crew chief, an assistant, and the flight line crew chief. Next were those responsible for individual systems, such as hydraulics, electrical, and powerplant. The third group of specialists looked after the Deuce's fire control and weapons systems. In addition, it was common practice for technical representatives from Convair to be available to troubleshoot problems.

The Air Force made a sweeping change in its maintenance structure by instituting Consolidated Aircraft Maintenance Squadrons, called CAMRONS, or CAMS. The units were comprised of maintenance personnel drawn primarily from air defense squadrons. The concept, which had originated in the SAC, placed CAMRONS directly subordinate to

An F-102A of the Interceptor Weapons School at Perrin AFB in 1959. (Larry Davis Collection)

An F-102A of the 482nd FIS, which was based at Seymour-Johnson AFB, North Carolina. During the time the unit operated Deuces—from 1959 to 1965—it deployed F-102 detachments throughout the southern states in response to crisis situations. (Terry Love Collection)

"Mr. Magoo II" (s/n 56-1103) belonged to the 86th FIS, based at Youngstown AFB, Ohio. Seen here in 1958, the Deuce was destroyed in a crash on 6 January 1960, one month before the unit turned in its last F-102. (Col. W. Gatschet via Marty Isham)

The 326th FIS used this tail marking, seen in 1959, to identify their F-102s. (Marty Isham Collection)

the group level. At bases where two squadrons were assigned, a single CAMRON, which typically had its own aircraft, was responsible for all maintenance activities, often of more than one type of aircraft. The consolidated maintenance program was viewed with disdain by pilots and ground crew alike, for a number of reasons. Morale suffered since crew chiefs, who identified with their aircraft, were no longer an integral part of the unit to which their aircraft was assigned. Gone was the camaraderie between the aircrews and the technicians who took care of their airplanes.

By the end of 1957, the ADC's self-evaluation of its defense capabilities was less than bright. Lengthy conversion schedules continued to have an adverse effect on air defense, and interceptors were still considered incapable of dealing with high-speed bombers above 40,000 feet. The Air Force synopsis held that the F-102 weapon system did not provide the ADC with an acceptable combat capability. Deuce pilots

Maj. Budd Butcher poses with his Deuce shortly before his unit, the 327th FIS, deployed to Thule AB in June 1958. (Budd Butcher Collection)

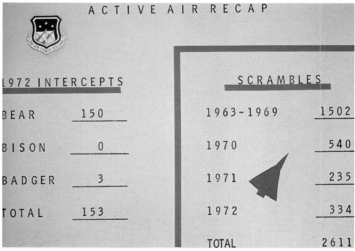

This photo of the Air Forces Iceland actual tally board best describes the command's activity levels during the Cold War. (Budd Butcher)

This F-102A of the 325th FIS wears nose markings indicating its participation in the massive movement of aircraft during the Cuban Missile Crisis in 1962. The rectangular placard below the markings, which was red and white, was required for any Air Force aircraft carrying live ordnance. Technicians are servicing the fire control system. (Author)

In 1972, special markings were added to drop tanks of 57th FIS F-102s, signifying the high incidence of encounters between the "Black Knights and the Russian "Bear." The knight's shield and rudder of the aircraft in the design match those of the unit's F-102s. (Kenneth W. Buchanan via Jerry Geer)

shared that view, though with more candor. They called unreliable Falcon missiles "Sand Seekers," cursed the failure-prone radar power equipment, and had even less confidence in rocket armament. In response to their mistrust of the Deuce as a weapons platform, some pilots, in the spirit of devout patriotism (and no doubt a bit of fighter pilot bravado), claimed that they would, if necessary, ram an enemy bomber to prevent it from reaching its target.

A problem with external fuel tanks proved detrimental to the Deuce's operating parameters. The ADC restricted the use of drop tanks due to a high incidence of unexplained tank releases. Subsequently, most missions were limited to an hour and a half, however, pilots added the tanks whenever possible to broaden the safety margin, especially in regions notorious for bad weather.

Despite drawbacks, those whose livelihood centered around the Deuce persevered. F-102 pilots professed a genuine love for the airplane and found it fun to fly. Crew chiefs seemed equally enthusiastic about its ease of maintenance. So, it became a matter of coping, pending the arrival of the F-101B and F-106A, still two years distant.

It was standard practice for F-102s of the Air Defense Command to wear their squadron emblem on the left side of the tail fin, while the ADC insignia appeared opposite. Deuces of the Iceland-based 57th FIS were easily identified by their checked rudders. During the 1960s, 11 rows of white and black, or blue, checks accented red tail fins. Beginning in 1970, the pattern was changed to larger, and subsequently fewer, checks in white and black. The pennant near the top of the fin was the ADC's "A" Award, which recognized units that distinguished themselves by sustained operational effectiveness. (USAF)

Following a crash on 26 February 1966 at Goose AB, Labrador, on 26 February 1966, this Deuce of the 59th FIS was repaired and placed back in service. (USAF)

Assigned to the 318th FIS, this Deuce wears the parent 325th Fighter Group emblem, a large yellow disc with black trim, on the tail fin and the squadron insignia on the nose. (Marty Isham Collection)

Meanwhile, reductions in the number of Air Force units continued as the American military machine kept pace with Russia's changing offensive posture. Waves of heavy bombers, the USSR's preferred method of attack, gave way to single, low-flying bombers with hydrogen bombs, ICBMs, and missile-laden submarines. The dubious effectiveness of manned interceptors against those threats loosened the very foundation on which the F-102's future was built.

Life for ADC pilots was much like that of firefighters— hours of boredom punctuated by intense, fast-moving episodes triggered by the sounding of an alarm. Throughout the globe, in Cold War hot spots where the two world powers played their cat and mouse, do or dare game, F-102s thun-

dered into the not-so-friendly skies to fulfill their design purpose: intercept, investigate, and, if necessary, annihilate.

The greater number of F-102 squadrons was concentrated along the United States' northern rim to guard against air attack via the polar air routes, while the second largest concentration was in the northeast regions, whose industrial centers and population density were surely highlighted on Russian plotting tables. The long arm of the ADC extended to Newfoundland, Greenland, and Iceland to guard extreme northern approaches.

Delta Dagger-equipped fighter interceptor squadrons at those locations didn't have to wait long for encounters with the Soviet bomber fleet. In the category for number of inter-

F-102s of the 57th FIS fly over Iceland during the 1970s. (USAF)

Between a pair of Russian "Bear" bombers, an F-102 of the 57th FIS takes a greater interest in the one equipped for mid-air refueling. The 57th intercepted more Soviet aircraft in 1970 than the preceding three years combined. (USAF)

Seen at a SAC base, this 64th FIS F-102A was apparently the victim of one of many tactical exercises in which Strategic Air Command units played the aggressor to test operational security. (Author)

cepts, the "Black Knights" of the Iceland-based 57th FIS easily took first place. Caught between the two superpowers, Iceland was at the edge of the northern shield. Up until 1970, Headquarters Air Forces Iceland recorded 1,502 scrambles. As Iceland's days became longer during early spring 1970, Norwegian radar detected increased air activity at airfields near Murmansk. The Soviet bomber fleet was making its presence known on a regular basis, probing NATO's radar shield, thereby providing 57th Deuces with a steady diet of scrambles. By years end, the Black Knights had racked up more than 500 scrambles, 347 of which were intercepts of Russian aircraft. Often, so many Russian aircraft were encountered that the radar station on Iceland's east coast, named "Critic," had its hands full assigning F-102s to individual intruders. Reflecting the heightened level of Cold War activity, on 15 September 1972 the 57th became the first in Air Force history to reach the 1,000 intercepts mark.

Newly assigned to the 37th FIS, this Deuce would soon exchange its former 498th FIS markings, which consisted of a red vertical band and "Geiger Tiger" emblem, for those of the 37th. The pair is seen at Ethan Allen AFB during October 1959. (USAF)

A pair of 57th Deuces flanks a Tu-20 "Bear" intercepted near Iceland in 1970. (USAF)

A Soviet Bear bomber flies with an unlikely wing man, an F-102A of the 57th FIS. Equipped for in-flight refueling, the Tu-20 could reach any major city in the U.S. (USAF)

The intercept mission itself required that F-102s fly extremely close to suspect aircraft to make accurate identification, take photographs, note details, and basically "fly the flag." One Deuce pilot came so close to a Soviet bomber to photograph what appeared to be a camera port that the Soviet Union filed a diplomatic protest. Similar missions were carried out by the 327th FIS, whose F-102s rose from Thule AB, Greenland, to meet Soviet aircraft over the North Atlantic and Arctic Ocean. Previously based at George AFB, California, the "Iron Mask" squadron traded in their original F-102s for those with Case XX wings and arctic markings in preparation for deployment to Thule. Since Greenland had been updated with new TACAN equipment, corresponding gear replaced the aircraft's original VOR system. That placed the Deuces ahead of the ground environment enroute to Thule, necessitating navigational assistance from F-89 Scorpions—not the best choice, since cruise speeds were vastly different. Thule Air Base, which had been built in 1952 in northwestern Greenland, was only 930 miles from the North Pole.

Cold weather scrambles varied somewhat from those in other climes. Most desirous were heated alert hangars near the runway. Aircraft parked in frigid air for a substantial period of time were totally dependent on ground support heating equipment. Missile temperatures could not fall below zero degrees Fahrenheit, nor could various fluids and sensitive components be subjected to extreme cold for extended periods. Wing and canopy covers, if available, were usually necessary, and extra caution was required to taxi, take off, and land on ice and snow. Extra effort was needed to strap in pilots wearing extreme weather clothing. Since white-outs and fog often obscured airfields, a number of methods were tried to make runways conspicuous, including spraying them with colored dyes and outlining them with tree branches. Deuce pilots were ever mindful of the chances of survival should their single engine fail over forbidding regions.

Not all intercept missions resulted in a positive identification. Every time that an ADC pilot was sent aloft, he experienced anticipation, perhaps even a twinge of anxiety, never

Deuces were usually scrambled in pairs, from takeoff to landing, such as this pair of 57th aircraft departing Keflavik, Iceland, during June 1965. (Baldur Sveinsson)

Prior to assignment to a fighter interceptor squadron, this F-102A served the 4750th Test Squadron. Here, it retracts the landing gear during take off from Tyndall AFB. A camera is mounted beneath the right wing. (USAF)

Two green stripes bordered by white on the aft fuselage identify this 48th FIS Deuce as the unit Operations Officer's aircraft. The pattern of green stars varied slightly from one aircraft to another. (Merle Olmsted via David McLaren)

sure of what he would encounter. Periodically, on the return track, he was still uncertain, since F-102 pilots were scrambled for UFOs. Radar sightings often turned out to be errant civil aircraft. The unexplained, however, were filed in the Air Force "Project Bluebook." On a much different note, southernmost Air Force and Air National Guard squadrons equipped with F-102s and F-106s received Presidential orders in 1972 to go on five-minute alert status as part of the strategy of the anti-drug smuggling campaign. Under "Operation Intercept," the interceptors were to work in concert with new low-level radar to target light planes that crossed southern U.S. borders loaded with illegal drugs.

Air Defense Command F-102s were key elements in what was perhaps the most serious and certainly the most involved confrontation of the Cold War. The Cuban Missile Crisis drew world attention when, in October 1962, the superpowers perched on the brink of nuclear war as President Kennedy

and Premier Khrushchev faced off over Soviet missiles and fighters secretly brought into Cuba. In managing the crisis, the Department of Defense took major steps to defend the Southeastern United States.

Prime radar installations at NAS Key West and Miami, Florida, were augmented by radar picket ship and Airborne Early Warning stations off Florida's southern tip. Aircraft committed to the effort included Navy fighters and F-102s from the 482nd FIS at Seymour Johnson AFB, North Carolina, and the Air Defense Weapons Center at Florida's Tyndall AFB. The use of F-102s to stay abreast of Cuba's Fidel Castro actually began during 1961. To counter possible hostilities by Castro against many of his detractors, who had taken refuge in the Miami area, six ADWC F-102s at Tyndall were repositioned south to Homestead AFB for alert duties. The "Southern Tip" exercise, which began on 12 April, was intended as a two-week trial, however, the decision was made to maintain a permanent vigil.

Members of the 332nd FIS "Full House" squadron pose with their F-102s at Tyndall AFB during Falcon missile training in 1958. A red chevron on the tail fin accented the unit emblem, which consisted of a "full house" in playing cards superimposed over crossed red Falcon missiles. (R. Johnson)

Besides TF-102As, a number of T-33s were kept in ADC squadron inventories to maintain pilot proficiency. This well-maintained "T-Bird" belonged to the 48th FIS at Langley AFB. (Author)

A pair of F-102As assigned to the 317th FIS fly over Alaska's Mount McKinley in April 1968. Soon after the unit transitioned to the F-102, it was transferred from the ADC to the Alaskan Air Command. The wing fences were painted yellow, and red lightning bolts were applied to the drop tanks. (USAF)

Deuces of the 482nd were substituted for the Tyndall aircraft in July, and by December 1961, the squadron had established a permanent alert detachment at Homestead. When the crisis flared in October 1962, the entire 482nd FIS deployed to Homestead, with some F-102s repositioned to the southernmost defensive site at NAS Key West in the Florida Straits. On 20 October, the 73rd Air Division directed nearly 60 of its aircraft (a mix of F-101, F-102, and F-106s) to five-minute alert status with external fuel tanks and armament, a portion of which was nuclear. Interceptors that stood continental alert with nuclear missiles were code named "Victor Alert." Backup defense was provided by Navy and Air Force fighters, which flew sorties from bases throughout the state

of Florida. By the time alert status had reverted to peacetime levels, a total of 33 F-102As and 16 TF-102As had been committed to the crisis.

Never completely dropping its guard, the DoD had the 482nd shift its alert detachment from Homestead AFB to NAS Key West during June 1963. The squadron ceased alert operations on 1 July 1965, just three months before it was inactivated. Recurring tensions with Cuba in 1967 warranted the attention of the NORAD, which directed the 32nd Air Division at Gunter AFB, Alabama, to assume interceptor alert at Key West. Accordingly, the ADWC established Alert Detachment 3 with eight F-102s. Aircraft were rotated from Tyndall with 4756th CCTS crews until U.S. Navy and TAC F-4s took over in November 1969.

Blue, or in some cases green, stars were applied over a white background to create the dazzling effect used by the 48th FIS to identify its F-102s. This Deuce is seen at Langley Field during May 1959. (Merle Olmsted via David McLaren)

A silver-painted Deuce of the 57th FIS shadows a Soviet Bear bomber. (USAF)

All 57th FIS markings were painted out of this F-102A, except for the Icelandic figure on the nose, prior to the aircraft's departure from Iceland when replacement F-4 Phantoms arrived. (USN)

Assigned to the 4756th ADG, this weathered F-102A wears two missile firing markers on its nose. The hapless Deuce was shot down by an F-106 in 1965. (Author)

Although training was a byword in the Air Defense Command, instruction was at first crude, unorganized, and followed no standard, but became better structured as more F-102s became available to squadrons. When the first ADC Delta Daggers arrived at George AFB, the 327th's Captain Martin O. Detlie was chosen to head the unit's training program in view of his exemplary flying record. Both Detlie and the squadron maintenance officer, Captain C.F. Davenport, trained with the F-102 at the Air Force Flight Test Center at Edwards AFB. Detlie, in turn, used the first Deuce delivered to the unit to train four key squadron officers.

Similar training was conducted on the opposite coast, where the 52nd Fighter Group at Suffolk County AFB received the first ADC Deuces in the east. The 52nd comprised the 2nd and 5th Fighter Interceptor Squadrons, whose mission was to protect New York City from air attack. Among the more than 50 F-102s, which began to arrive on 14 January 1957, were three TF-102As, which each pilot flew twice as a stu-

dent prior to 30 hours of combat interception training in the single-seaters. That was followed by two or three training missions per day for those who qualified. Before they were considered combat ready, pilots were required to complete a number of profiles, which encompassed all phases of an intercept mission. "Chase" missions were also flown in which an instructor followed a new pilot through every flight step of combat readiness training. Instrument flights, which were required for proficiency ratings, were flown in T-33 or TF-102A trainers.

Since the Air Force had no specialized F-102 school, training by individual squadrons became commonplace. Typically, a few squadron pilots, usually the most experienced, were selected for Convair's ground school at San Diego. After five check rides, they ferried new Deuces back to their units, where transition programs with TF-102As were set up and, if a unit was fortunate to have one, an MD-3 F-102 simulator. The first few solo flights were usually limited to VFR conditions,

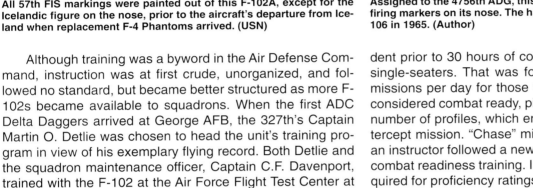

In an impressive display of power and beauty, a Deuce of the 57th FIS touches down at Keflavik, Iceland. (USAF)

The checked pattern on the rudder, which also appeared in blue and white colors, was worn by Deuces of the 57th FIS during the 1960s. International Orange was also used as a high-visibility color, which, in the 1970s, was applied only to the tail fin and wing tips. (Author)

followed by flights with ceiling and visibility restrictions. It was not unusual for instructor pilots, who required only 25 hours in the type for that rating, to check out pilots from other units. This pattern continued until 1960, when F-102s were assigned to the Air Training Command at Perrin AFB, Texas, where they shared ramp space with F-86 Sabres used for training ADC pilots.

Since its activation in 1941, Perrin was largely responsible for turning out single-engine pilots, and in 1952 began training interceptor aircrews. Since the ballistic missile threat loomed so ominously by 1961 that the ADC was forced to place one-third of its force on alert status, it became apparent that not enough pilots were combat ready. Also obvious was the lack of standardization in training among fighter interceptor squadrons. Although F-102 training began at Perrin during 1960, the time had come to intensify the training necessary to put fully qualified interceptor pilots in the cockpit.

The first step in reaching that goal was the ADC's acquisition of Perrin AFB in mid-1962 from the Air Training Command. The base and its 4780th Air Defense Wing (Training) were assigned to the 73rd AD, which was headquartered at Tyndall, also home to the 73rd's 4756th Air Defense Wing (Training) and Interceptor Weapons School, which operated 20 TF-102As and 15 F-102As. Subordinate to the 4780th ADW was the 4780th Air Base Group, which governed the 4781st and 4782nd Combat Crew Training Squadrons. With those units, the IWS, and newly formed F-101 and F-106 combat crew training schools, the 73rd AD became solely responsible for all ADC interceptor pilot qualification. The ATC at Perrin had already laid the groundwork for transitioning pilots to the F-102, which the 73rd AD enhanced with an academics syllabus and actual mission profiles in an air defense environment.

Training in the ADC had finally become commensurate with mission requirements. During its heyday, the school's 100 pilots that staffed the 73rd's two squadrons turned out 120 new combat-ready F-102 pilots per year, plus 50 previously qualified jet pilots put through an accelerated F-102 program. Still others completed lead-in training for added instruction in other types. In anticipation of foreign F-102 deliveries, "Project Peace Violet" got under way at Perrin during November 1967 for the training of six Turkish instructor pilots. Under the same program, six Greek Air Force IPs began training on 30 September 1968.

The course itself lasted 26 weeks and combined 130 classroom hours, 147 flying hours (93 sorties), and 293 hours of briefing and simulator time. In the interest of pilot safety, the ADC's Life Support School was established at Lake

With the move of the 64th FIS from McChord AFB to Paine Field, Washington, in March 1960 came new tail markings. Three of the four flights within the squadron are represented by the color of each aircraft's speed boards. (Marty Isham Collection)

Texoma, near Perrin AFB, where survival and ejection seat training was conducted for ADC aircrew and Air National Guard units assigned the ADC mission. Training climaxed with live firing at drone targets over Tyndall's Gulf range. Pilots then flew back to Perrin for graduation. All F-102 training at Perrin ceased by July 1971 with the 111th FIS of the Texas ANG assuming all F-102 training. With the phaseout of the F-102, plus the opening of the Dallas-Fort Worth International Airport, Perrin AFB itself was closed during 1972.

Within the ADC was the 4750th Test Squadron, which operated and evaluated all Air Defense Command fighter aircraft. The unit was activated on 1 September 1956 at Vincent AFB, near Yuma, Arizona, and had four F-102s permanently assigned. The squadron was transferred to the 73rd Air Division during July 1957 and moved to Tyndall AFB, where it came under the 4756th Air Defense Group (Weapons) in July 1961. In January 1963, its control was shifted to the 4756th Air Defense Wing (Training), and in January 1968, it was reassigned to the Air Defense Weapons Center. The ADWC had been established on 31 October 1967 at Tyndall, where it inherited the Interceptor Weapons School and its fleet of F-102s. Every ADC squadron rotated through the ADWC for air defense and weapons training, until it was transferred to the Tactical Air Command on 1 October 1979.

Besides a strict training regimen, ADC squadrons frequently engaged in exercises held to hone not only their skills, but those of participating units outside of the air defense community, including other services and NATO air arms. It was not uncommon for air defense exercises to span nation-wide and last for months at a time. In the late 1950s, during a period of mounting tension over the Formosa Crisis, North American air defenses faced their most severe test as the Strategic Air Command launched one of its largest-scale mock attacks without warning. Throughout the continent, U.S. and Canadian lines of defense, including DEW lines, picket ships, radar installations, and interceptors countered the SAC bombers, many of which approached from the North Pole.

During an exercise that lasted from October 1961 to January 1962, called "Project Dry Martini," F-102s of the 326th FIS were pitted against SAC B-58 "Hustler" bombers in intercept trials. During mid-1964, F-102s teamed up with EC-121D Airborne Early Warning platforms in the Canal Zone for "Exercise Cashew Tree." In addition, F-102 squadrons became regular participants in the "William Tell" worldwide weapons competitions, beginning in 1958.

Although operational activities and busy training schedules took an expected toll on aircraft, the F-102's accident rate was high compared to that of other tactical aircraft. Pilots conceded that the Deuce possessed some quirks that could get them in trouble, most notably, the delta-wing configuration, which produced gradual stalls and subsequently, barely noticeable sink rates. However, the majority of F-102 losses in the Air Defense Command were attributed to engine problems, with landing gear failure a close second. The first loss occurred on 25 February 1957 when a 317th FIS

Deuce suffered engine failure and crashed at McChord AFB.

From that date to 22 January 1973, the ADC recorded 119 F-102As destroyed, a dozen of which were TF-102As, among a total of 219 accidents. Included in those grim statistics are nine destroyed in mid-air collisions and two shot down by USAF aircraft: one by a GAR missile in 1959, and another downed by an F-106 in 1965.

In a strange case of turnabout, the only manned U.S. aircraft shot down by an F-102 was a T-33 of the 328th Fighter Group on 22 August 1958. The 328th at Richards-Gebaur AFB was hosting the 11th FIS while their home base at Duluth underwent runway repair. While on a practice intercept mission with the "T-Bird," the Deuce inadvertently followed through with the firing sequence when it closed the distance. When the T-33 crew saw the F-102's missile bay doors snap open, they bailed out, just seconds before their aircraft was destroyed. It is doubtful that the 11th FIS ever considered applying a T-33 silhouette below the F-102's cockpit, at least not while with their host unit.

During its peak year, the type alone suffered 62 accidents, 28 of which resulted in the aircraft's destruction. In stark contrast, fate smiled on the busy ADC for only a brief period—the month of October 1961, the command's first accident-free month for all aircraft types. As testimony to the reduction in aircraft, steady training, and expert flying skills, the years 1971 and '72 were accident-free for ADC F-102s.

A number of factors contributed to the F-102's state of flux during the time it served the Air Defense Command as a first-line interceptor. In conjunction with ongoing modification programs, Deuces were constantly rotated through squadrons, which typically had 25 to 28 aircraft assigned. As updated models arrived, older aircraft were put through the IRAN process and passed to other commands to bolster their capabilities. In some cases, entire squadrons were relocated to ensure maximum air defense coverage of global interests. The 317th and 31st were sent to Alaska in 1957, followed shortly thereafter by the 323rd's move to Newfoundland. In 1958, Greenland welcomed the 327th, which was replaced by the 332nd in 1960.

As the last active Air Force squadron to fly the Deuce, the 57th FIS transitioned to the F-4 Phantom during 1973. (Baldur Sveinsson)

During the transition period, the first F-4 assigned to the 57th FIS accompanies an F-102 on its final mission. (USAF)

European and Pacific commands began acquiring Deuces from the ADC, beginning in 1959. That year also marked the arrival of the F-101B and F-106A, which figured in to the steady decline in the number of F-102s assigned to the ADC. In 1960, the first Air National Guard F-102s arrived in Texas. By July, the number of ADC F-102 squadrons had dropped from 27 to 15. By year's end, only 9 squadrons remained, causing the accumulation of 160 surplus Delta Daggers at ADC bases. During the mid-1960s, Deuces were pulled out of the ADC inventory for commitments in the Pacific and the war in Southeast Asia. The ADC's total number of fighter squadrons had fallen to 35 by the end of June 1966. From that point on, the USAF contribution to the NORAD was substantially reduced. All F-102 squadrons, along with nine F-101 squadrons, were slated for inactivation by 1970.

By 1968, the 57th FIS was the only F-102 fighter interceptor squadron remaining in the ADC. When a 57th Deuce crashed on 22 January 1973, it marked the final loss of an F-102 in the Air Defense Command. The loss brought the unit down to 13 aircraft, which began departing Iceland in pairs on 2 May 1973, ending the Deuce's service as an operational ADC interceptor. Throughout its career in the Air Defense Command, the F-102 had filled the ranks of 33 fighter interceptor squadrons.

The following are brief organizational histories of the Air Defense Command fighter interceptor squadrons that operated the F-102:

2nd FIS - Originally assigned to the 4709th Defense Wing with F-86Ds at McGuire AFB, New Jersey, the 2nd was relocated to Suffolk County AFB, New Jersey, with its transfer to the 52nd Fighter Group (Air Defense) on 18 August 1955 in accordance with "Project Arrow," which was a major reorganization of Continental Air Defense squadrons. The 2nd converted to F-102s in January 1957, to be replaced by F-101Bs beginning in December 1959. The squadron's last F-102 departed Suffolk County during early February 1960.

5th FIS - Like its sister squadron, the 2nd FIS, the 5th was moved from McGuire AFB to Suffolk County AFB on 18 August 1955 with its transfer to the 52nd FG (AD). The unit exchanged its F-86Ds for F-102s beginning in March 1957. It was transferred to the 32nd FG (AD) and moved to Minot AFB, North Dakota, with its conversion to F-106As during February 1960. The 5th FIS flew the Deuce until March.

11th FIS - Based at the municipal airport at Duluth, the 11th exchanged its F-89Ds for F-102s beginning August 1956, following its transfer from the 515th ADG to the 343rd FG (AD). It began converting to the F-106A in July 1960.

18th FIS - Flying the F-89D, the 18th left the Alaskan Air Command in August 1957 for assignment to the 412th FG at Wurtsmith AFB, Michigan, with F-102s. Before the "Blue Foxes" had completely exchanged its F-102s for F-101Bs by mid-August 1960, it was transferred to the 478th FG (AD) and moved to Grand Forks AFB, North Dakota, on 1 May 1960.

27th FIS - Based at Griffiss AFB, New York, its F-94Cs were replaced by F-102s beginning in early June 1957. Before the 27th "Falcons" had fully converted to the F-106A during February 1960, the squadron was relocated to Loring AFB, Maine, during October 1959.

31st FIS - Having been inactivated on 18 August 1955 as part of "Project Arrow," the 31st was reactivated on 8 June 1956 at Wurtsmith AFB, Michigan, and assigned to the 412th FG (AD) with F-102s. The first F-102s didn't arrive until 9 December. When the squadron was finally up to full strength by February 1957, it shared the Wurtsmith facility with F-89Js of the 445th FIS. In keeping with Air Force policy of rotating units without affecting air defense coverage, the 31st traded places with the 18th FIS at Ladd AFB, Alaska, beginning in October 1957.

37th FIS - Based at Ethan Allen AFB, Vermont, under the 14th FG, the 37th began converting from the F-86D to F-102 during December 1957. The unit was inactivated on 1 May 1960.

47th FIS - Just six months after it had converted to its third version of the Sabre, the F-86L, the 47th began transitioning to the F-102 in March 1958. Based at Niagara Falls Municipal Airport, New York, the squadron was inactivated on 1 July 1960.

48th FIS - Assigned to the 85th AD, the 48th began trading its F-94Cs for F-102s in March 1957. Although it remained at Langley AFB, Virginia, the unit came under the Washington ADS on 1 September 1958. By October 1960, it had fully converted to the F-106A.

57th FIS - During the time that it flew F-89Cs from Keflavik Airport, Iceland, the "Black Knights" of the 57th was the only fighter unit in the Military Air Transport Service, since Keflavik was a MATS base. When the responsibility for Keflavik went to the U.S. Navy, the 57th was reassigned to the ADC on 1 July 1962, equipped with F-102s, and reported to Air Forces Iceland. Small by Air Force standards, the squadron never had more than 14 aircraft assigned. The 57th was the last active USAF unit to relinquish the F-102, having flown them until July 1973, when they were replaced by F-4Cs.

Like the F-102s they replaced, F-4C Phantoms of the 57th FIS carried on the tradition of sporting black and white checked tail fins. (John Guillen)

59th FIS - When the F-102 began replacing the unit's F-89Js during May 1960 at Goose AB, Labrador, the 59th "Black Bats" was transferred from the 4732nd ADG to the Goose ADS. By the end of 1966, the squadron had relinquished its F-102s and was inactivated shortly thereafter.

61st FIS - With its transfer from Ernest Harmon AFB, Newfoundland, to Truax Field, Wisconsin, under the 327th FG (AD) during October 1957, the 61st exchanged its F-89Ds for F-102s. By July the following year, the squadron had turned in its aircraft and the unit was inactivated during July 1960.

64th FIS - When transferred from the AAC to the ADC and assigned to McChord AFB, Washington, under the 325th FG (AD) on 15 August 1957, the 64th received F-102s. On 15 March 1960, it was transferred to the 326th FG (AD) and relocated to Paine Field, Washington. Another transfer on 1 April 1961 placed it under the 57th FG (AD), and on 6 June 1966 the 64th was reassigned to the PACAF.

71st FIS - Having previously flown the F-86L, the 71st converted to the F-102 beginning in October 1958. The squadron, based at Selfridge AFB, Michigan, under the 1st FG (AD), transitioned to the F-106A in October 1960.

76th FIS - Shortly after its transfer from McCoy AFB, Florida, to Westover AFB, Massachusetts, the unit turned in its F-89Js for F-102s beginning in March 1961. The last F-102 left the 76th during March 1963 and it was inactivated in July.

82nd FIS - Based at Travis AFB, California, the squadron made the switch from F-86Ds to F-102s during early July 1957. Although remaining at Travis, successive transfers to the San Francisco ADS, Portland ADS, and 26th AD culminated with its reassignment and relocation to the PACAF on 25 June 1966.

86th FIS - Based at Youngstown AFB, Ohio, since it was activated, the 86th converted from the F-86D to the F-102 beginning in August 1957. The last F-102 left the unit on 4 February 1960, and it was inactivated on 1 March.

87th FIS - Shortly after the 87th transferred from the 58th AD to the 30th AD on 1 September 1958, the squadron began receiving F-102s to replace its F-86Ls at Lockbourne AFB, Ohio. Its last F-102 departed Lockbourne on 22 June 1960, when the conversion was made to F-106As.

95th FIS - Having flown F-86Ds under the 85th AD from Andrews AFB, Maryland, the 95th began switching over to F-102s in early January 1958. Although conversion to the F-106A began in September 1959, the last F-102 remained until August 1960.

317th FIS - Having replaced F-86Ds with F-102s at McChord AFB, Washington, beginning in early December 1956, the 317th soon transferred from the ADC's 325th FG (AD) to the AAC in December 1957.

318th FIS - The squadron and its F-86Ds was transferred from the 525th ADG and relocated from Presque Isle AFB, Maine, to McChord AFB, Washington, under the 325th FG as a sister unit to the 317th to bolster northeast air defense. The conversion to F-102s began during early January 1957. During March 1960, its transfer to the 325th FW (AD) initiated the conversion to the F-106A, which was completed by May.

323rd FIS - One of a number of units affected by "Project Arrow" in 1955, the 323rd, which reported to the 327th FG, moved to Truax Field, Wisconsin, with its F-86Ds, converting to the F-102 in November 1956. One year later, it was transferred to the 4731st ADG and moved to Ernest Harmon AFB Newfoundland. Within two weeks after transfer to the Goose ADS on 1 June 1960, the 323rd relinquished its F-102s, and it was inactivated on 1 July.

325th FIS - After transferring from the 566th ADG to the 327th FG (AD) and moved to Truax Field, Wisconsin, with its F-86Ds to replace the 456th FIS under "Project Arrow," the 325th converted to the F-102 in February 1957. The unit outlasted other F-102 squadrons previously assigned to Truax, operating the Deuce until its inactivation on 25 June 1966.

326th FIS - At the time that Grandview AFB, Missouri, was renamed Richards-Gebaur AFB, during April 1957, the 326th "Skywolves" began exchanging its F-86Ds for F-102s, which it flew until the unit's inactivation on 2 January 1967.

327th FIS - When the "Iron Mask" squadron was reactivated under "Project Arrow" on 18 August 1955, pilots of the famed 94th FIS "Hat in the Ring," flying F-86Ds, were transferred as a group to the 327th under the 27th AD. Their job, as the first squadron in the ADC to receive the F-102, was to put the Deuce through its operational paces. The first F-102 arrived at George AFB, California, on 24 April 1956. The 327th was the first F-102-equipped fighter interceptor squadron to go abroad, deploying to Thule AB, Greenland, during June 1958. It officially became part of the 65th AD on 3 July 1958 and was inactivated on 25 March 1960.

329th FIS - Assigned as the sister unit to the 327th under "Project Arrow," the 329th replaced its F-86Ls with F-102s beginning in September 1958 with its transfer from the 4722nd ADG to the 27th AD. It was transferred to the Los Angeles ADS on 1 October 1959, and during October 1960, the first of its F-106A replacements arrived at George AFB.

331st FIS - When the unit was transferred from the 33rd AD to the Albuquerque ADS during January 1960, its conversion from the F-86L to the F-102 began. After transfer to the Oklahoma City ADS, the 4752nd ADW and back to Oklahoma City, the conversion to F-104A/Bs began in April 1963.

332nd FIS - Based at McGuire AFB, New Jersey, as part of the 4730th ADC, the 332nd "Full House" squadron began acquiring F-102s to replace its F-86D/Ls in July 1957. The squadron was transferred to the 33rd AD on 1 July 1959 and moved to England AFB, Louisiana. It was transferred to the Oklahoma City ADS on 1 January 1960, and during July, began its deployment to Thule AB to relieve the 327th FIS. It came under the 4683rd ADW on 1 September 1960 until it was inactivated on 1 July 1965.

438th FIS - When it was transferred from the 534th ADG to the 507th FG (AD) on 18 August 1955, the 438th operated F-89Ds, which were replaced by F-102s, which began to arrive at Kinross AFB, Michigan, during March 1957. The base was renamed Kincheloe AFB during late 1959. Conversion to the F-106A began during June 1960.

456th FIS - Originally based at Truax Field, Wisconsin, until replaced by the 325th FIS, the 456th was reactivated on 18 October 1955 at Castle AFB, California, and assigned to the 28th AD flying first F-86D, then F-86L Sabres. Those were replaced by F-102s in April 1958, and F-102As first arrived to take their place in September 1959.

460th FIS - F-102s began to arrive at Portland Airport during January 1958 to replace the unit's F-89Ds. The 460th operated its F-102s under the 337th FG until F-106As took their place in February 1966.

482nd FIS - Assigned to the 85th AD and based at Seymour-Johnson AFB, North Carolina, the 482nd began exchanging their F-86Ds for F-102s during April 1957. The squadron changed parent commands with transfers to the 325th AD, 32nd AD, and Washington ADS before it was inactivated on 10 October 1965.

498th FIS - Activated under "Project Arrow" in August 1955, the 498th was assigned to the 84th FG (AD) flying F-86Ds from Geiger Field, Washington. The "Geiger Tigers" began converting to the F-102 in January 1957 until all were replaced by F-106As by October 1959. The following year, Geiger Field became Spokane International Airport.

Total F-102s Assigned to ADC

DATE	INVENTORY
30 June 1956	5
31 December 1956	97
30 June 1957	300
31 December 1957	428
30 June 1958	495
31 December 1958	651
30 June 1959	621
31 December 1959	482
30 June 1960	281
31 December 1960	212
30 June 1961	193
31 December 1961	226
30 June 1962	221
31 December 1962	228
30 June 1963	176
31 December 1963	194
30 June 1964	191
31 December 1964	193
30 June 1965	193
31 December 1965	160
30 June 1966	59
31 December 1966	11
30 June 1967	14
31 December 1967	14
30 June 1968	13
31 December 1968	21
30 June 1969	12
31 December 1969	14
30 June 1970	14
31 December 1970	14
30 June 1971	14
31 December 1971	14
30 June 1972	14
31 December 1972	14

Beginning in 1959, the F-106A "Delta Dart" replaced the F-102 as the premiere air defense interceptor. (NASA/Tony Landis)

Top Cover for America

The importance of establishing Alaska as part of America's military defense systems was first fully realized during the post-World War I period. And no one person realized the strategic importance of the northern territory more than outspoken Brigadier General William L. Mitchell. His profound advocacy of Alaska's position as the "Air Crossroads of the future" eventually altered the military's perception of the barren region's strategic significance. General Henry "Hap" Arnold, who shared Mitchell's views, ordered the first flying squadrons to Alaska in 1940. Thus began a steady infusion of military units, which culminated in the war over the Aleutians.

After World War II, the Joint Chiefs of Staff determined that every nation capable of going to war with the U.S. was located north of the 45th Parallel. That their most direct attack routes were over the Arctic Ocean pointed out the need for an air defense system across Canada and Alaska. Subsequently, an elaborate three-part plan was put into motion. An aircraft control and warning system became fully functional during 1954, and was augmented by a vast network of distant early warning radar sites which became known as the "DEW Line." Both systems were linked by an equally extensive communications system called "White Alice." Completing the air defense triad were air units equipped with interceptor aircraft.

During 1957, when Alaska's air defense was placed under control of the NORAD, the air defense force of the Alaskan Air Command (AAC) had peaked with six squadrons of F-89 Scorpion interceptors. That same year, Alaska's air defense system was divided into two sectors, with the northern half's control center located at Ladd AFB and the southern portion's at Elmendorf AFB. The year also marked the beginning of draw-downs in view of the Soviet switch from bombers to intercontinental ballistic missiles as its primary strategic weapon. In conjunction with plans to counter with Alaska-based long-range missile sites, the AAC pressed for F-102s to replace its F-89s. It was obvious that acquisition of the superior Deuce would pay huge dividends in operational capabilities and the reduction of fighter interceptor squadrons. The number of squadrons was trimmed from six to three during late 1957, and the F-102 was introduced to Alaska, its first assignment to another Air Force command.

On 15 August, the 317th FIS, with 25 F-102As and 4 TF-102As, officially transferred from the ADC, and on 1 September 1957 the first two F-102s departed McChord AFB, Washington, for their new home at Elmendorf AFB, Alaska. The second Deuce squadron assigned to the AAC was the 31st FIS, based at Wurtsmith AFB, Michigan. It was officially transferred on 20 August, and the first of its 23 F-102s began arriving at Elmendorf on 16 September. The Scorpion-equipped 64th, 65th, and 66th Fighter Interceptor Squadrons departed as more Deuces arrived.

Meanwhile, at Ladd AFB in the 11th Air Division, F-89J conversion kits were arriving for the 449th FIS. The 18th and 433rd Fighter Interceptor Squadrons at Ladd were reassigned, leaving only the 449th. Further reductions resulted in the inactivation of the 31st FIS on 8 October 1958 just one year after it began operations. Its F-102s went to the 31st, which then boasted 46 Deuces, the largest number operated by a squadron. That left only two fighter interceptor squadrons in Alaska: the 449th and 317th, the latter of which operated alert detachments at Eielson AFB, Ladd AFB, and forward operating bases at King Salmon and Galena. The last Scorpions were phased out of the AAC by the end of July 1960, leaving complete responsibility for the air defense of Alaska and northern approaches to the F-102s.

The Deuce's arrival in Alaska stirred Soviet air reconnaissance activity beginning in March 1958. More than three years after an unsuccessful intercept attempt in September 1958, two 317th pilots, who scrambled from Galena, intercepted a pair of Tu-16 Badger bombers over the Bering Sea,

Trimmed with Insignia Red, a fully modified F-102A of the 317th FIS is prepared for a mission from Elmendorf AFB. (USAF)

A pair of colorfully marked 317th Deuces flies near Elmendorf during a Tactical Air Command cold weather exercise in March 1969. (USAF)

A 317th FIS Deuce at the point of touchdown at Elmendorf AFB during August 1969. The emblem of the parent 21st Composite Wing was worn on the left side of the tail fin. (Norm Taylor)

off Alaska's northwest coast, on 5 December 1961. Unfortunately, the next intercept mission would not meet with the same favorable results. On 15 March 1963, a Soviet Tu-22 Blinder overflew Nunivak Island and Alaska's western coast. A pair of Deuces scrambled from King Salmon, however, due to their limited range, were forced to turn back just 20 miles from the Soviet aircraft. State department and political repercussions over the Soviet violation of Alaska's airspace and the F-102's inability to complete the intercept brought about a number of proposals. Foremost concerned the acquisition of F-4C Phantoms, however, those efforts were thwarted, forcing the AAC leadership to agree to rotational deployments of ADC F-106s.

The first nine Delta Darts arrived at Elmendorf AFB in July 1963 under the code name "White Shoes." Throughout the decade, Soviet bombers periodically flew near Alaskan territory, undoubtedly to "tickle" the American radar network and monitor aircraft response. One example of the cat and mouse games played by both superpowers during the Cold War, which illustrates the danger of such tactics, occurred during the 1962 Cuban Missile Crisis. It is likely that the most tense day of the crisis was 27 October, when a SAC U-2 spy plane, which staged from Eielson AFB, accidentally flew into Soviet airspace. By the time the U-2 pilot realized his error, Soviet Mig interceptors had been scrambled to shoot him down. The U-2 pilot made radio contact with a U.S. command post in Alaska, which directed him to steer due east and quickly return to U.S. airspace. Meanwhile, 317th FIS F-102s, armed with nuclear Falcon missiles, were scrambled to protect the U-2. That placed the decision whether to use

Four TF-102As were assigned to the 317th FIS, twice the number normally carried on a squadron inventory. Throughout the unit's history, it operated a total of 44 Deuces before it was inactivated. (James Wogstad Collection)

High-visibility red tail sections were the norm for aircraft operating in Alaska's snowy regions. The red areas covering the aft fuselage were later repainted the aircraft color. Here, a flight of four from the 317th FIS sweeps across the Alaskan skies during February 1958. (USAF)

F-102As, some of which are adorned with native Alaskan figures on their noses, stand alert duty at Galena Air Force Station during September 1960. The Deuces were assigned to the 317th FIS. (Marty Isham Collection)

nuclear weapons, during a most anxious time, in the hands of Deuce pilots. In that particular episode, time was on everyone's side and disaster was averted.

The Soviets often sent more than one aircraft. In February 1968, three bombers flew to within 80 miles of Alaska, and on 4 April 1969, a flight of 8 to 10 Badgers closed to within 60 miles of the Alaskan coastline before they were intercepted by F-102s scrambled from Eielson AFB. The F-106s made 17 successful intercepts, while Deuces of the 317th accomplished 13 during their Alaskan duty. From 1961 to 1969, Alaskan Air Command F-102s completed a total of 15 intercepts of 23 Soviet aircraft, most of which occurred over the Bering Sea and involved Deuces launched from Galena. Russian aircraft weren't the only subjects photographed by 317th pilots. After a disastrous earthquake struck Alaska on 27 March 1964, TF-102As, with Air Force photographers

aboard, provided photographs that enabled officials to plan recovery efforts.

On 25 August 1960 the 317th FIS was placed under the 5040th Air Base Wing, where it remained until 8 July 1966, when it was inactivated and all Elmendorf flying units were assigned to the 21st Composite Wing. On 24 November 1965, the 317th was awarded the Hughes Trophy for outstanding performance of the fighter intercept mission. To its credit, the unit received the combat proficiency award on two later occasions.

Throughout its existence, the 317th's status, like the Alaskan Air Command itself, remained in jeopardy as air defense doctrine and the perception of Alaska's strategic importance changed. Both were examples of the steady decline in Alaska's military forces, which was aggravated by the growing conflict in Southeast Asia. By mid-1969, the 317th's F-

When all of the Alaskan Air Command's flying units were placed under the 21st Composite Wing in 1966, F-102s, EB-57Es, and T-33As were grouped in the 317th FIS. Two of the squadron's aircraft are seen here near the top of Mount McKinley during March 1969. (USAF)

Liberal amounts of red paint were used to make F-102s of the 317th FIS more visible against Alaska's barren terrain. The Deuce in the right background wears two white stars on a blue field near the top of the tail fin. (Marty Isham Collection)

An F-102A of the 317th at Elmendorf AFB on 17 May 1969. An Air Force Outstanding Unit Award is displayed below the cockpit. (Norm Taylor)

A 317th Deuce at Alaska's Elmendorf AFB during May 1964. (Norm Taylor)

Nose landing gear failure plagued the F-102 throughout its service career, and Deuces that operated in cold climates were more susceptible to the problem. This 317th Deuce ended up on its nose at Elmendorf AFB during March 1969. (USAF)

102 inventory had been reduced to 27 aircraft. Rotational alert deployment of F-106s, which had been renamed "College Shoes," continued. The Alaskan Air Command was hard hit by the 1969 budget constraints, which spelled the end of the 317th FIS. The Alaskan Air Command suffered the loss of 14 F-102s of the 317th from crashes.

Following the loss of a Deuce during June, the unit had 26 F-102s slated for transfer. Five left Elmendorf for Ellington AFB, Texas, on 10 December, and one was removed from the inventory for donation to the Alaska Transportation Museum. The remaining 20 were flown to Truax Field, Wisconsin, on 12 December for assignment to the Wisconsin Air National Guard. The 317th was officially inactivated on the 31st, leaving the AAC with a void in its air defense capabilities. On 23 June 1970, the 43rd Tactical Fighter Squadron arrived in Alaska with its F-4E Phantoms to assume the air defense role. Although the Phantoms were superior performers and gave the AAC a tactical advantage, it would never be the same as when the familiar red-tailed Deuces ruled the Alaskan skies.

The Deuce in Europe

A key element in upholding America's vow to defend Europe against communist aggression during the Cold War was the United States Air Forces in Europe (USAFE). The command was a vital component in fulfilling North Atlantic Treaty Organization commitments. USAFE was under control of both the U.S. Air Force and the U.S. European Command. Its forces within central and northern Europe were also elements of the Supreme Headquarters Allied Powers Europe (SHAPE), which was responsible for all military assets of the nations that comprised Air Forces Central Europe (AFCENT). Had there been a major confrontation with the Soviet Union or any of the Warsaw Pact countries, USAFE would have come under direct control of NATO's Allied Air Forces Central Europe (AAFCE).

The AAFCE was divided into the 2nd and 4th Allied Tactical Air Forces (ATAF). Under the 4th ATAF, the 17th Air Force at Ramstein Air Base, West Germany, governed USAF-NATO unit operations within France, Germany, the Netherlands, Italy, Belgium, Norway, and Libya. As part of a massive, cooperative effort, air units of those nations, along with those of the Royal Air Force, worked regularly with USAFE air elements.

USAFE components in Spain were not incorporated into AFCENT, but were subordinate to the 65th Air Division, which served as sentinel over American interests in Spain. In response to a request from the Spanish government, the 65th AD was established under the Strategic Air Command's 16th Air Force at Torrejon AB on 8 April 1957. The 16th AF was part of NATO's Allied Forces Southern Europe (AFSOUTH), which comprised the air arms of Spain, Italy, Greece, and Turkey.

A total of six F-102 squadrons were entrusted with the air defense of central, northern, and western Europe. As NATO units, they were under operational control of the 86th Air Division at Ramstein AB, which originated as the 86th Fighter Interceptor Wing. It was placed under the command of the 17th Air Force on 15 November 1959 and was redesignated as an air division on 18 November 1960.

The fighter interceptor squadrons guarded the airspace over central Germany, known as Air Defense Sector 3. One squadron based in the Netherlands was under the operational control of the Royal Netherlands Air Force as part of the 2nd ATAF. It covered the airspace over the Netherlands, northern Germany, and portions of the North Sea, known as Air Defense Sector 1. Coverage of the largest area, which stretched from Spain to Turkey, for which the 16th AF was responsible, was disproportionate with only two interceptor squadrons assigned. Both squadrons, along with the 65th AD, were transferred from SAC to USAFE on 1 July 1960. Although technically based in Europe, the 57th FIS at NAS Keflavik, Iceland, remained under ADC control.

In light of the fact that U.S.-based ADC units had begun exchanging their early model Deuces for upgraded versions, the decision was made during late 1958 to send F-102s to USAFE as replacements for F-86s. A great deal of preparation was necessary prior to surface shipment of the F-102s. A total of 164 Deuces (150 F-102As and 14 TF-102As) were involved in the original deployment. All, with the exception of 18 (which were ferried from locations outside of the continental U.S.), were processed through three Air Material Area (AMA) depots as part of the Preparation for Overseas Transfer Program. The San Antonio AMA at Kelly AFB, Texas, received 93 F-102s, the San Bernardino AMA at Norton AFB, California, prepared 44, and 9 were processed through the Ogden AMA at Hill AFB, Utah.

To ease logistics, an effort was made to select F-102s for overseas assignment from the same production blocks. As

Sixteen F-102s of the 525th FIS form an impressive display for an inspection at Germany's Bitburg Air Base during the early 1960s. The 525th was the first fighter interceptor squadron in Europe to transition to the Deuce. (USAF)

A long way from home, an F-102A of the 59th FIS is kept company by an F-105D on the return trip from Spain's CASA depot in March 1962. (R. Satterfield via Marty Isham)

Seen here in June 1959, while assigned to the 6250th Test Group at Hanscom AFB, Massachusetts, and the Cambridge Research Center, number 845 was one of two Deuces attached to the 526th FIS in Germany from February 1963 to June 1964 for trials with the new data link Air Weapons and Control System. The wing fences and tail fin tip were red with white stars. (Jim Burridge)

Having just been transferred from the 317th to the 525th FIS, this Deuce had the new unit markings applied over those of its former unit. (David McLaren Collection)

F-102s were released from ADC squadrons beginning in 1958, they were flown to the AMAs, where they underwent thorough inspections and various upgrades. A number of F-102s ended up in storage for periods up to five months as a result of the Air Force's indecisiveness in releasing aircraft from the ADC and assigning them to squadrons.

Modifications at the AMAs included a TACAN installation, external fuel tanks, a pilot survival kit, and an updated MG-10 AWCS. The F-102s also received new paint jobs for both aesthetics and preservation. The complete time that an aircraft spent in the Preparation for Overseas Program ranged from 60 to 80 days. When completed, the Deuces were passed to the Mobile AMA at Brookley AFB, Alabama, where they came under the Tactical Air Command's 4440th Aircraft Delivery Group. After cocooning (which, in true Air Force tradition, received its own code name—"Project Seaspray"), the Deuces were loaded aboard U.S. Navy aircraft carriers for the Atlantic crossing. After the two-week voyage, the carriers

arrived at Saint Nazaire, France, where the Deuces were off loaded and towed to the nearby Sud-Aviation facility at Montoir Air Base. There, the cocooning material was removed, the aircraft tested, and ferried to their new assignments.

The first group of F-102s arrived at the Mobile embarkation point during December 1958 and docked at the French port on 9 January 1959. Ten more crossings were made, with the final shipment arriving during October 1960. On 28 January 1959, the 525th FIS at Bitburg AB, Germany, received five Deuces (two of which were trainers), becoming the first Europe-based unit to transition to the type. The Delta Dagger was formally introduced on the continent at an air show in February. The last of 25 Deuces touched down at Bitburg on 7 March, and on 1 July the squadron became operational, standing "Zulu Alert" for NATO.

The 496th FIS at Hahn AB, Germany, received its first four F-102s on 9 December 1959. Next came the 497th FIS at Torrejon AB, Spain, which received its first Delta Dagger

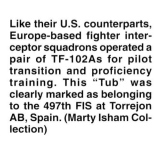

Like their U.S. counterparts, Europe-based fighter interceptor squadrons operated a pair of TF-102As for pilot transition and proficiency training. This "Tub" was clearly marked as belonging to the 497th FIS at Torrejon AB, Spain. (Marty Isham Collection)

Armed with Falcon missiles, an early model F-102A of the 431st FIS "Red Devils" stands alert at Zaragoza AB, Spain. Both the 431st and 497th FIS were assigned to the 65th Air Division and bolstered Spain's air defense. (Author)

An F-102A of the 32nd FIS based at Soesterberg AB, Holland, in 1966. The 56th AD insignia was typically worn on the right side of the tail fin, while the squadron emblem was applied to the left side. (Neal Schneider)

on 28 September. The first three F-102s for the 32nd FIS landed at Wheelus AB, Libya, on 12 August 1960. Although the 32nd was based at Soesterberg AB, Holland, its pilots were not qualified in the Deuce and underwent training at Wheelus.

The point at which the F-102 reached full operational status in Europe proved timely in view of the construction of the Berlin Wall, which began during August 1961. During the tense period, Europe-based F-102 squadrons conducted operations with three F-102 and three F-104 squadrons of the Air National Guard, which were among 11 squadrons called to active duty and deployed to Europe.

Due to their proximity to communist forces, the F-102 units in Europe maintained a keen state of readiness. Normal procedure had a minimum of one flight of F-102s in the air over Europe throughout most of a 24-hour period. In addition, each squadron kept two Deuces on 5-minute alert, with

another pair on 15-minute alert. Another two were kept on one-hour standby. That alert posture was relaxed somewhat on 3 November 1964 with two aircraft remaining on 5-minute alert, while eight were placed on six-hour standby. Pilots usually worked an eight-day duty cycle, which comprised two days of day alert, two of night alert, two days of training, and two days of rest. Like their stateside counterparts, pilots and ground crews lived in four-bay "Zulu" hangars throughout their duty period. In the event of a scramble, a pair of Deuces could be airborne in less than four minutes and, if necessary, could be at 40,000 feet in less than five minutes after takeoff.

For tactical purposes, NATO had divided Central Europe into four Air Defense Sectors, each of which had its own operations center. Although not under NATO control, units based in Spain operated under a similar system. Like those in the U.S., the fighter interceptor squadrons were linked to a network of ground radar stations as part of a vast communica-

A 32nd FIS Deuce in commander's stripes at the 1965 Paris Air Show. This aircraft was transferred to the 525th FIS during early 1964, but was returned later that year in view of its fitting serial number. The lone 86th AD emblem was the second style of tail markings used by the unit. (Jean Magendie via Stephen Miller)

The colorful markings familiar to USAF aircraft gave way to the "tri-tone" camouflage born of necessity for Southeast Asia. The drab scheme, introduced to USAFE Deuces during 1965, did little to enhance the F-102's sleek silhouette. (Denis Hughes via Nick Williams)

tions and control system. The Deuces were tasked primarily with intercepting unidentified aircraft detected in a 30-mile-wide corridor between East and West airspace. Europe-based Deuces amassed flight hours quickly during weekly scrambles into the zone to visually identify aircraft, which were usually discovered to be errant airliners. Occasionally, communist military aircraft strayed into the zone, but were quick to return to their airspace. Similar duties had the Deuces scrambled aloft to escort aircraft in trouble. On numerous occasions, F-102s assisted pilots who experienced mechanical or navigational problems.

On missions that promised higher anxiety levels, F-102s were often involved in the well-established, yet little-known cat and mouse game of "tickling" Soviet and Warsaw Pact air defense systems. On such forays, groups of F-102s flew at near maximum speed and altitude toward communist borders, breaking off the run just short of the hostile boundaries. Certain that their airspace was being penetrated, communist radar was activated to track the intruders, allowing U.S. intel-

ligence gathering aircraft to monitor and evaluate their response.

Delta Dagger pilots and ground crewmen maintained a high degree of proficiency through a rigorous training schedule. Throughout their tours in Europe, pilots were rotated through Wheelus AB, Libya, where they sharpened their skills on air-to-air ranges over the Mediterranean. Wheelus, near Tripoli on the North African continent, was home to the 7272nd Air Base Group of the Europe Weapons Center, which conducted F-102 advanced weapons system training. Its 7235th Support Squadron was equipped with Martin B-57Es modified for target towing. The squadron also operated F-100Cs, which flew as control aircraft during F-102 training missions. Beginning in 1960, each Deuce squadron was supplied with inert Falcon missiles for ongoing training. Called the "Weapons System Evaluation Missile," the dummy weapon never left its firing rail, but electronically and photographically recorded all stages of a missile firing sequence. Beginning in October 1967, EB-57Es of the 4713th Defense Systems

Distinctive among the tail markings worn by USAFE Deuces was that used to identify aircraft of the 431st FIS. The emblem was designed by a 431st crewman who also painstakingly applied many of the emblems. (Jack M. Friell)

The red, white, and blue fin markings familiar to the 32nd FIS were applied during the unit's initial training at Wheelus AB. The unique squadron crest was applied later. (Jack M. Friell)

This colorful Deuce served as the 497th FIS commander's mount during 1964. The blue, yellow, and red diagonal stripes were continued on to the fuel tanks. A heart with the words "Sweet Lips" appears below the anti-glare panel. (Jack M. Friell)

Evaluation Squadron arrived to conduct training in electronic countermeasures previously carried out by Douglas RB-66s, which had been pulled out of Europe for the war in Southeast Asia.

When it became apparent that Libya would oust U.S. forces, F-102 training was diverted to Zaragoza AB, Spain, during February 1970. Ongoing training was also accomplished by deployments to Spain's Torrejon AB. Each fighter interceptor squadron used Lockheed T-33s for target and support duties. When the Deuce arrived, the "T-Birds" remained, however, those based in Germany were combined with F-102s of the 526th FIS and pilots stayed proficient in both types.

The abilities of USAFE Deuce crews were evaluated during participation in the William Tell Worldwide Weapons Meet, held every two years at Florida's Tyndall AFB. Closer to their European home bases, F-102s played a major part in the annual AFCENT Air Defense Competition, first held in 1965. The month-long event was based on testing the ability of NATO aircraft to safeguard their particular air defense sector. A wide range of scramble, intercept, and attack missions were scored, along with ground control and weapon loading events. Keen on competitive spirit, USAFE leaders organized "LOADEO," an annual Munitions Loading Competition, also begun in 1965. Munitions loading crews in four aircraft categories were scored on technique, professionalism, equipment, and a written examination.

To avoid the logistical complications of returning F-102s to the U.S. for model upgrades and overhaul, such services were accomplished in Europe under civil contract or by Air Force personnel. Progressive modifications of the F-102 fleet spanned the entire period that Deuces were in Europe. Beginning in spring 1960, F-102s of the 496th and 525th squadrons received upgrades and new paint jobs at the Fiat facili-

Seen at Soelingen AB, Germany, in December 1964, this F-102A was assigned to the 526th FIS based at Ramstein AB. The tail band and the lightning bolt on the fuel tank were red. (Author)

Deuces of the 526th FIS wore a red band on the tail fin with superimposed squadron and 86th AD emblems from 1964 to 1966. This well-maintained F-102A is seen at RAF Lakenheath during July 1966. (Marty Isham Collection)

An F-102A of the 525th "Bulldogs" at Sculthorpe, Great Britain, in May 1962. The striking blue and white tail design was used by the unit from 1961 to 1963. The red rectangle immediately aft of the national insignia registered onboard munitions data. (Jack M. Friell)

The commander's aircraft of the 526th FIS taxis at Soesterberg AB during March 1965. (J.D. Ragay via Robert F. Dorr)

Europe-based F-102s underwent gunnery training with Martin B-57Es of the 7235th Support Squadron based at Wheelus AB, Libya. The B-57Es were modified with target-towing equipment beneath the rear fuselage. The radar-reflective target was reeled out 20,000 feet over Mediterranean ranges. (USAF)

Details of the 525th's F-102 tail markings used from November 1964 to October 1966. The ribbon below the serial number is the Air Force Outstanding Unit Award. (Jerry Geer)

ties at Turin-Caselle, Italy. Additional modifications began during 1961 at the Breguet plant, Chateauroux-Deols, France. Prior to the Deuce's arrival in Europe, the French facility had been established as the central depot for the Air Material Force European Area (AMFEA) under partnership with the San Antonio AMA. After successful trials of an arrestor hook-equipped F-102, hooks were installed on Deuces of the 86th Air Division beginning in April 1961. The Chateauroux facility performed the modification to 65th AD F-102s during 1962.

The responsibility for depot-level maintenance was turned over to the CASA Aircraft Company at Seville-San Pablo, Spain. CASA began contract work during April 1962, and by September 1969 had overhauled 411 F-102s. Portions of the program that involved the installation of classified radar and electronics were carried out by Hughes Aircraft Company technicians. Upgrades to the Figure-8 level were completed during 1964, the same year that Deuce engines were overhauled by CASA, under "Project Blowtorch."

Modernization of Central Europe's air defense radar stations began during 1963 with a new system that directed missions by data link instead of radio transmissions. By the time the system had become fully operational in early 1965, the installation of associated equipment in the aircraft had begun.

The USAFE F-102 fleet was brought up to Figure-10 standards during 1966, and in 1967 further improvements were made to engines, ejection seats, and external fuel tanks, plus anti-collision lights were added. Before the Deuce was phased out of Europe, it underwent modifications to its UHF radio system.

The operational activities of USAFE squadrons were based primarily on NATO air operations within host countries and assigned air defense sectors. Practice was the norm as Deuces flew sorties against aircraft of other units at their home base and those of other NATO air arms. The 32nd FIS was unique in that it was the only USAF air unit controlled by a

A fully upgraded TF-102A of the 32nd FIS. The "Tub" was used by squadrons for pilot training and instrument test flights. (V. Vandenberg)

A freshly camouflaged Deuce of the 525th FIS during 1966. The aircraft was destroyed in a crash two years later. (Jack M. Friell)

foreign government. Its assignment to the Royal Netherlands Air Force served as a momentous gesture of goodwill between the U.S. and NATO members. The squadron conducted operations with Hawker Hunter aircraft of the RNAF's No. 325 Squadron, also based at Soesterberg. Further promoting relations between the U.S. and her allies was the joint use of alert hangars by both American and Dutch aircraft beginning in 1963. Aside from normal operations, the 32nd FIS deployed six aircraft to Erding AB, Germany, in January 1967, when a Soviet Mig-17 landed in Bavaria. Also noteworthy was the squadron's selection to represent USAFE in the 1965 William Tell competition, in which it placed first.

The 431st FIS and 497th FIS, assigned to the 65th Air Division, were the USAF's contribution to Spain's air defense system. Both units shared operational duties with F-86F Sabres of the Spanish Air Force's 21st Fighter Squadron. Of major importance was their protection of SAC B-47 bombers regularly deployed to Zaragoza AB as part of "Project REFLEX." For a brief period in 1960, the 497th had the distinction of flying the only F-102s assigned to the Strategic Air Command, which occurred prior to the unit's transfer to USAFE to consolidate Europe's fighter assets.

As the first of three squadrons based in Germany to transition to the F-102, the 525th FIS spearheaded many of the training phases necessary for F-102 operations in Europe. A 525th pilot made the first arrested landing with an F-102 on 23 July 1963 at Bitburg AB. The 526th FIS at Ramstein AB played a key role in the establishment of the data link Air Weapons and Control System, which was vital to Europe's air defense radar net. Under the code name "GRAY GHOST," 526th pilots flew half of nearly 240 sorties during 1963, which were necessary to fully evaluate and implement the data link system. The sorties were flown in conjunction with two specially-equipped JEF-102As assigned to the unit for the trials.

When SAC discontinued B-47 bomber rotations through Spain's Zaragoza AB in 1964, the 431st and 497th squadrons were slated for return to the U.S. and transfer to the Tactical Air Command. Since the F-102s of the 496th were early models, 20 newer aircraft of the departing 497th were transferred, with the remaining Spain-based Deuces passed to other squadrons in Europe. The older types were flown back to the U.S., along with 23 from other USAFE units under "Project KRAZY KAT" during April and June 1964. Under "Project HARDWAY I," an additional 7 Deuces were pulled out of Europe during September 1964, with another 5 withdrawn in December under "Project HARDWAY II."

Since 1964, USAFE leaders deemed the F-102 inadequate in comparison to its communist adversaries. Since its intended replacement, the F-4 Phantom, was heavily committed to the war in Southeast Asia, plans to phase the Deuce out of Europe met long delays. Meanwhile, the only pressing need for F-102 defense coverage beyond normal operating parameters occurred when France withdrew from NATO in 1966. The move required that a six-aircraft F-102 detachment be established at Erding AB, Germany, which was manned on a rotational basis by 86th AD squadrons. As a cost-saving measure and to consolidate all fighter and interceptor units, the three fighter interceptor squadrons in Germany were placed under control of the resident tactical wing at their respective base, while the 32nd FIS came under 17th Air Force control. Included in the November 1968 realignment was the 496th's assignment to the 50th Tactical Fighter Wing at Hahn AB, Germany, the 525th's to the 36th TFW at Bitburg AB, and the 526th's to the 26th Tactical Reconnaissance Wing at Ramstein AB. The 32nd FIS became a tactical fighter squadron (air defense) on 1 July 1969, and its first replacement F-4E arrived the following month. The three squadrons based in Germany followed suit with the 525th redesignated on 1 October 1969, the 496th on 1 January 1970, and the 526th on 1 April 1970. As additional F-4Es arrived in Europe, the F-102s were flown back to the U.S. for service with the Air National Guard. The 526th flew the last USAFE F-102 mission on 1 April 1970, ending the Delta Dagger's career in Europe. In more than a decade of air defense duty in Europe, a total of 21 F-102s were lost due to operational causes.

All F-102s arrived in Europe in the standard overall gray finish, which, in many cases, had to be reapplied after arrival due to damage caused by the removal of cocooning material. Deuces also underwent respray during overhaul and modification phases. Shortly after arrival at their squadrons, F-102s sported colorful markings, which were typical of USAFE aircraft during that period. The F-102's prominent tail fin served as the canvas to which was applied bold, colorful patterns of wide bands and sharp-angled stripes. Unit emblems were usually superimposed over the tail colors, the exception being the 497th FIS, which displayed only the emblem. After 1963, the lone emblem became the standard for all F-102s based in Europe. Standard practice had the squadron emblem applied to the left side of the fin, while the division insignia appeared on the right.

At the same time that camouflage was being considered for USAF aircraft in Southeast Asia, the decision was made in 1965 to apply the scheme to USAFE Deuces during scheduled overhaul and modification phases. In the meantime, USAFE interceptor squadrons reverted to adding color to the tails of their F-102s in the form of wide, colored bands. Once camouflage was applied, the aircraft remained void of squadron and personal markings. During the colorful years, F-102s flown by unit commanders wore colored stripes around the fuselage. These were usually red or multi-colored to represent each flight's identifying color. The 526th FIS operated more than one Deuce with "commander's stripes" since the unit hosted high-ranking staff pilots who maintained proficiency with 526th aircraft.

Many of the markings worn by F-102s, other than those on the tail, were based on a number of factors, including squadron policy, unit location, or personal preference. Such was the case with the 32nd FIS in Holland, many of whose aircraft had a pair of wooden shoes painted on the nose. Shortly after their arrival, the 496th, 525th, and 526th squadrons applied F-102 silhouettes to the nose of their aircraft, which served as backgrounds for pilot and crew chief titles. Since squadrons were divided into four flights, each section's identifying color was often displayed somewhere on the aircraft. The components most often painted were the braking parachute housing (speed boards), intake boundary layer slabs, external fuel tanks, and landing gear doors.

USAFE F-I02 UNITS

SQUADRON	LOCATION	A/C ARRIVAL	OPERATIONAL	INACTIVATED/REDES.
32nd FIS	Soesterberg AB, Neth.	8-12-60	2-9-61	7-1-69
431st FIS	Zaragoza AB, Spain	9-28-60	10-24-60	3-1-64
496th FIS	Hahn AB, Germany	12-9-59	5-23-60	1-1-70
497th FIS	Torrejon AB, Spain	9-14-60	6-18-64	10-1-69
525th FIS	Bitburg AB, Germany	4-26-60	1-28-59	6-7-60
526th FIS	Ramstein AB, Germany	11-15-60	4-1-70	7-1-59

An F-102A of the 32nd FIS based at Soesterberg AB, Holland, near the Dutch city of Utrecht. Seen in June 1966, the Deuce wears a rectangular emblem on the air intake below the cockpit, which denotes the squadron 's first place position in the 1965 William Tell Weapons Meet. (Author)

Delta Daggers in the Far East

Vital to the U.S. Air Force commitment to protect American interests in eastern Asia was its Pacific Air Forces (PACAF), formerly known as the Far East Air Forces (FEAF), which kept watch on the skies from northern Russia to China's northern regions.

Despite such awesome responsibilities, PACAF initially governed only six fighter interceptor squadrons, which were tactically dispersed throughout the Pacific. During the late 1950s, three were positioned in Japan, two in Okinawa, and one at Guam, equipped primarily with F-86D Sabres. As new interceptor types were brought into the Air Defense Command, F-102s became available for assignment to PACAF units. Shortly after the F-102's introduction to those units, PACAF would become extensively involved with the war in Southeast Asia, dramatically changing the face of interceptor operations in the Pacific.

The first Deuces in the Pacific went to the 16th FIS at Naha AB, Okinawa, in March 1959. The following month, the 509th FIS, which had previously flown F-100s, began operations at Clark AB in the Philippines as an interceptor squadron with F-102s. As part of the simultaneous changes at Clark, the 509th's parent command became the 405th Fighter Wing, which replaced the 6200th Air Base Wing. After the 509th was fully operational with the Deuce, two PACAF F-86D squadrons were inactivated: the 41st FIS at Andersen AB, Guam, on 8 March 1960; and the 25th FIS at Naha AB on 8 June. The 68th FIS at Itazuke AB, Japan, took delivery of their first F-102s during December 1959, followed shortly thereafter by the 4th FIS at Misawa AB and the 40th at Yokota AB, Japan.

Those five squadrons operated the F-102 within PACAF for nearly five years. During that time, detachments from the 16th and 509th Fighter Interceptor Squadrons were rotated throughout Southeast Asia for air defense and varied combat duties, with the 509th bearing the brunt of combat assignments. Other activities conducted by PACAF Deuce units included escort flights, training sorties, and competitions with other wings throughout PACAF air defense sectors.

When Air Force leaders were certain that F-102s could be phased out of PACAF, despite the growing war machine, the decision was made to reorganize PACAF interceptor squadrons. The 68th FIS was redesignated as a Tactical Fighter Squadron on 24 July 1964 and began transitioning to the F-4C Phantom; the 16th FIS did likewise beginning on 24 December. The 4th and 40th Fighter Interceptor Squadrons followed suit with redesignations to TFS on 20 June 1965, and changeovers to Phantom models. By late 1965, the 509th FIS operated the only PACAF F-102s. With the phasing out of Deuce units and subsequent aircraft transfers, plus the introduction of IR seeker-equipped Deuces during late 1964, the 509th eventually ended up with all fully modified F-102s. Earlier models withdrawn from PACAF were returned to the U.S. for assignment to the Air National Guard. By July 1965, the 509th boasted a fleet of 42 F-102s (four of which were TF-102As) to fulfill its wartime commitment.

Before long, the 509th found itself hard-pressed to supply F-102s for detachments throughout Southeast Asia. Attrition had reduced its Deuce inventory to 33 aircraft, 24 of which were necessary to equip detachments. To ease the situation, the decision was made to transfer F-102 units from the ADC to PACAF.

A pair of F-102As from the first production batch and assigned to the 16th FIS fly near their home base at Naha AB, Okinawa, during the early 1960s. The 16th was the first in the Pacific to acquire the Delta Dagger. (USAF)

Snow is cleared from Delta Daggers of the 4th FIS at Misawa AB, Japan, during February 1961. (Bill Thomas)

Five early model F-102As of the 68th FIS undergo maintenance at Itazuke AB, Japan. (General Dynamics)

Displaying its "Crusader" emblems on the nose and tail, this Deuce of the 68th FIS begins to show wear in January 1961. (Bill Thomas)

The first step occurred on 31 August 1965, when the Air Defense Command was directed to relocate an F-102 squadron to Naha AB to replace the Phantom-equipped 555th TFS. Although ADC's commander preferred to rotate units to PACAF, even though that adversely affected the ADC dispersal program and air defense of the U.S., the decision remained firm. The 82nd FIS at Travis AFB, California, was selected for the move to Naha.

The original plan was to transport the 82nd's F-102s by aircraft carrier, the method used to move all Deuces previously transferred to PACAF bases. During mid-September 1965, a proposal to fly the aircraft across the Pacific was presented as a viable means of completing the transfer. Obviously, a plan to refuel the F-102s in flight gave rise to a number of complexities, not the least of which involved sheer logistics and the implementation of an inflight refueling system. Although the idea was brushed aside and plans were

made for cocooning the aircraft prior to shipment, a number of delays and indecision in the Defense Secretary's office gave some plausibility to the idea of a Pacific flight.

Meanwhile, the 82nd was removed from alert status on 1 November with the movement of its 28 F-102s, code named Project "DEUCES WILD," slated to take place two weeks later. On 3 November, the Department of Defense approved not only the deployment, but gave the nod to proceed with the plan to mid-air refuel the F-102s across the Pacific, thus ending the dilemma over the method of transfer. The 82nd FIS was to be operationally ready at Naha on 25 February 1966.

The ADC saw this as an opportunity to give a large number of its F-102s refueling capability, thereby increasing their operational latitude as part of a worldwide deployment plan. A total of 87 F-102s were modified with air-to-air refueling capability. The modification itself, which was designed and developed by Convair, comprised an extended probe which

A Deuce of the 40th FIS at Yokota AB, Japan, in 1961. The bold insignia worn on the tails of 40th aircraft comprised a red devil's head and yellow lightning bolt over a white and red disc. (Marty Isham Collection)

A line-up of 12 Deuces of the 40th FIS at Yokota AB. The F-102A in the foreground wears the inscription "Road Runner II Beep Beep," while the second Deuce was named "Porcoff II." (Jerry Geer Collection)

A PACAF Deuce keeps close company with a 21st TFW F-100D over Japan. The Deuce is easily identified as a 4th FIS aircraft by red and black checks bordered by black trim. This Deuce takes its inscribed name, "Red-Striped Rascal," visible on the tail fin, from the four red angled stripes, also edged in black, on the forward fuselage, signifying the squadron commander's aircraft. (USAF)

Wearing the familiar checkerboard pattern on its tail, a Deuce of the 16th FIS stands alert at Naha AB. (USAF)

An F-102A of the 460th FIS shortly after its arrival at Sacramento ALC for modification for Project DEUCES WILD and assignment to the 82nd FIS. The aircraft would undergo overhaul, which included the installation of aerial refueling equipment and camouflage paint. (US Army)

An F-102A of the 64th FIS modified for aerial refueling during May 1966. (Marty Isham Collection)

ran the length of the starboard side of the fuselage. The probe receptacle was positioned next to and slightly above the cockpit. Since the probe hindered the Deuce's maximum flight performance, they were removed following the Transpacific flight.

The modification process was initially carried out by General Dynamics-Convair and the Air Force Logistics Command at San Diego, California, beginning on 4 October 1965. By year's end, 27 F-102s of the 82nd FIS had the system installed. Three additional ADC squadrons, the 64th, 325th, and 326th, began arriving at the AFLC depot at McClellan AFB, California, in January 1966 for the installation. A total of 86 Deuces, including two TF-102As, had the refueling package installed by June. One trainer variant was modified during fall 1967 as a replacement for a 509th aircraft. During the modification process, the Deuces also received the tri-tone camouflage scheme and updated liquid oxygen systems.

Visible on the fuselage behind the cockpit of this F-102A of the 4780th ADW are six mounts for an air-to-air refueling probe. The Deuce had previously been assigned to the 326th FIS, whose F-102s received the modification during 1966 before the unit was inactivated. (Neal Schneider)

The "Boomer's" view of an F-102A modified for aerial refueling prior to the Pacific deployment. The modified Deuce underwent mid-air refueling trials with a KC-97L of the Texas Air National Guard's 181st Air Refueling Squadron during 1965. (Texas ANG)

One of 87 F-102s modified for mid-air refueling is inspected by officials in March 1966. The awkward equipment proved detrimental to the aircraft's flight performance and was removed shortly after deployment. (Mitch Mayborn via Jerry Geer)

To prepare the 82nd air crews for the Pacific crossing, inflight refueling training was provided by the Tactical Air Command and Strategic Air Command. The ADC declared the 82nd FIS combat ready on 7 February 1966. Under the new code name Project "THIRSTY CAMEL," the first F-102 departed Travis AFB on 14 February. After stops at Hickam AFB, Hawaii, and Wake Island, the 82nd FIS arrived at Naha AB on 18 February. Four days later, ahead of schedule, the unit was standing alert duty. On 25 June, the 82nd officially became part of the PACAF. Meanwhile, units with modified F-102s that remained stateside practiced air-to-air refueling on a steady basis with KC-97 and KC-135 tanker aircraft.

During late February 1966, it was decided to boost the PACAF's F-102 inventory with two additional ADC squadrons. The 64th FIS at Paine Field, Washington, was selected to deploy to Clark AB for reassignment to PACAF, and the 325th FIS at Truax Field, Wisconsin, was to transfer to Misawa AB, Japan. In typical Air Force fashion, code names were assigned—the move to Clark was called Project HOT SPICE, and the move to Misawa became Project TALL TALE. The 64th's deployment was approved. However, the Department of Defense had the Air Force reconsider the 325th's relocation. Although the number of F-102s passed to the PACAF would not have diminished ADC's combat capability, since all Deuce units were slated for inactivation by mid-1968, it would have adversely affected plans to upgrade the Air National Guard's fleet of F-102s. Therefore, on 31 May 1966 the 325th's deployment was disapproved.

Flying from Hamilton AFB, California, the 64th FIS accomplished their Transpacific flight from 6 to 11 June 1966. Between the 82nd and 64th squadrons, a total of 52 F-102s had been successfully transferred. As PACAF F-102s were lost, replacement aircraft were drawn from the 4780th ADW, which had acquired F-102s modified for inflight refueling when the 325th and 326th were inactivated.

The Philippine-based 509th FIS and 64th FIS drew the majority of combat assignments in Southeast Asia, while the Naha-based 82nd FIS conducted operations outside the war zone. When the USS *PUEBLO* was captured by North Korea on 23 January 1968, the U.S. response included an alert detachment of the 82nd FIS, which deployed to Suwon, South Korea, during the crisis. The 64th FIS rotated their F-102s through Suwon as an added measure.

During alert duty at Suwon AB, Korea, during the Pueblo Crisis in 1968, Deuces of the 64th and 509th FIS fly formation with a South Korean flight demonstration team. (USAF)

A pair of Deuces are refueled from a KC-135E tanker enroute to Clark AB, Philippine Islands, during Operation HOT SPICE. (USAF)

The number of F-102s in the Pacific Air Forces peaked during mid-1969 with 84 aircraft. A total of five Deuces were lost during operations outside of the war zone. On 15 December 1969, the 64th was inactivated at Clark AB, with six of its Deuces turned over to the 82nd and 509th—the remainder were scrapped. Expanded F-4 operations in Southeast Asia enabled the 509th FIS to steadily reduce its operations during spring 1970 until it was finally inactivated on 24 July. That left the 82nd FIS as the last active duty F-102 squadron in the Pacific Air Forces. On 31 May 1971 it too was inactivated, and 35 of its Deuces were scrapped at Naha AB.

PACAF F-102 SQUADRONS

UNIT	LOCATION	OPERATIONAL DATES
4th FIS	Misawa AB, Japan	April 1960 - 6-20-65
16th FIS	Naha AB, Okinawa	3-9-59 - 12-24-64
40th FIS	Yokota AB, Japan	November 1960 - 6-20-65
64th FIS	Clark AB, P.I.	6-10-66 - 12-15-69
68th FIS	Itazuke AB, Japan	December 1959 - 7-24-64
82nd FIS	Naha AB, Okinawa	2-18-66 - 5-31-71
509th FIS	Clark AB, P.I.	April 1959 - 7-24-70

PACAF F-102 OPERATIONAL LOSSES

SERIAL NUMBER	UNIT	DATE LOST
55-3372	16th FIS	1-28-63
56-1146	509th FIS	1-31-67
56-1318	64th FIS	1-31-67
57-0893	82nd FIS	4-21-67
53-1791	509th FIS	9-7-67

F-102s MODIFIED FOR AIR-TO-AIR REFUELING
MODIFIED AT McCLELLAN FACILITY
56-1318 56-1328 56-1332 56-1333 56-1335 56-1338 56-1342
56-1343 56-1346 56-1347 56-1349 56-1352 56-1361 56-1362
56-1372 56-1382 56-1384 56-1385 56-1389 56-1391 56-1398
56-1420 56-1426 56-1436 56-1439 56-1444 56-1446 56-1449
56-1450 56-1451 56-1463 56-1464 56-1467 56-1469 56-1472
56-1490 56-1491 56-1493 56-1497 56-1499 56-1506 56-1509
56-1515 56-1516 56-1517 57-0787 57-0801 57-0807 57-0808
57-0812 57-0821 57-0829 57-0864 57-0878 57-0880 57-0909
56-2362 56-2366

MODIFIED AT SAN DIEGO FACILITY
56-1397 56-1440 56-1470 56-1507 57-0772 57-0774 57-0778
57-0779 57-0780 57-0783 57-0784 57-0794 57-0796 57-0799
57-0802 57-0804 57-0815 57-0840 57-0848 57-0851 57-0865
57-0882 57-0884 57-0887 57-0888 57-0891 57-0893 57-0895
56-2373

War in Southeast Asia

The only Delta Daggers to serve in a combat theater were from PACAF squadrons, which served rotational tours of duty at bases in South Vietnam and Thailand throughout most of the war in Southeast Asia. Assignments were so sporadic that it became difficult to track when and where Deuce contingents were actually in the war zone.

Although commonly thought to have begun wartime operations in 1962, the F-102 actually made its first appearance in Southeast Asia during 1961. During August, four F-102s of the 509th FIS at Clark Air Base were temporarily deployed to Don Muang Airport outside Bangkok, Thailand, to provide air defense. They relieved a flight of six F-100 Super Sabers from the 510th Tactical Fighter Squadron at Clark, which were deployed during April under the code name "Bell Tone" as part of President Kennedy's Air Force mission in Thailand.

Delta Daggers maintained their status as early participants in the conflict with a deployment of 509th FIS aircraft during 1962. What seemed to be part of an increase in communist activity in South Vietnam appeared on radar at Pleiku in the central regions. On the evening of 9 March, Pleiku radar showed seven unidentified flight tracks over the central highlands. A B-26 "Invader," scrambled from Bien Hoa Air Base and guided by ground control intercept (GCI) controllers to the area, found nothing of the intruders. Radar picked up similar low-level tracks coming out of Cambodia on the night of the 20th, but again, no aircraft were observed. Not surprisingly, many believed that the signals were contrived

as leverage to support the argument for deploying more military aid to the region.

Nevertheless, certain that the enemy was conducting overflights, the U.S. government dispatched F-102s to confront the unidentified aircraft. On 22 March, three F-102As and one TF-102A of the 509th FIS were sent from Clark AB to South Vietnam's Tan Son Nhut AB. The deployment was code named "Operation Water Glass," under which the Joint Chiefs of Staff authorized the Deuce pilots to destroy any hostile aircraft encountered over South Vietnam.

A bonafide fear of enemy air attack was prevalent among both political and military officials. Such apprehension may have been fueled by an attack on South Vietnam President Diem's palace on 26 February 1962 by a pair of mutinous South Vietnamese pilots in AD-6 Skyraiders. Looming over the deteriorating situation was the fear that Peking would step into the fray with Ilyushin Il-28 "Beagle" bombers. Realistically, though China and Vietnam had been enemies for centuries, the Chinese government seemed basically disinterested in the war to the south. The real threat was from Il-28s known to have been based at Hanoi in North Vietnam.

The F-102s flew countless practice GCI missions, and in July, began alternating six-week tours of air defense duty with Navy EA-1F early warning, all-weather Skyraiders. The unidentified radar signals dissipated as quickly as they had begun. After months of no signs of enemy air activity, General Harkins of the Military Assistance Command, Vietnam (MACV) proclaimed, "There is no air battle in Vietnam and there are

Equipped with air refueling probes, the first F-102s of the 82nd FIS flown to Naha AB, Okinawa, are seen during a stopover at Wake Island during April 1966. (Glen Sutton via Jerry Geer)

"Hawk Nest II" was an F-102A of the 509th FIS at Udorn RTAFB during December 1966. (Joe Sutila via Jerry Geer)

A pair of 64th FIS Deuces, which were temporarily assigned from Clark AB, stand alert at DaNang AB, South Vietnam, during May 1968. (Robert Mikesh)

no indications that one will develop." With that, plus severe overcrowding at Tan Son Nhut AB, air defense rotation at the base was brought to a temporary halt. The 509th FIS pulled all of its assets back to Clark in May 1963, however, the 13th Air Force conducted "no-notice" deployments of F-102s from Clark to South Vietnam until mid 1964. In addition, training flights were regularly flown to Tan Son Nhut and DaNang Air Bases. In response to unrest in the South Vietnamese military and political infrastructure during 1963, and mindful of the February 1962 coup attempt, the American government sent three F-102s to Tan Son Nhut during October. Their movement was in conjunction with the positioning of a U.S. naval task force off Vietnam. Neither was needed, however, since the U.S. government stayed clear of the coup, which began on I November.

By the end of 1963, U.S. military leaders remained confident that complete withdrawal of the Deuce from Southeast Asia could proceed as originally planned. The 68th FIS was the PACAF's first squadron to relinquish its F-102s, begin-

ning in July 1964. Next would be the 16th FIS, slated for inactivation on 24 December. However, actions by the North Vietnamese and U.S. political hierarchies brought about an abrupt change in course. At a time when "Vietnam" had yet to become a household word, events took place in the Tonkin Gulf during August 1964 that drew worldwide attention to the burgeoning conflict in Southeast Asia.

While on intelligence-gathering patrol off the North Vietnam coast on 2 August, two U.S. destroyers were attacked by communist PT boats. Although it is unlikely that a similar incident, claimed by the President to have taken place on 4 August, actually occurred, the incidents provided the Johnson administration with the leverage needed to apply more force to prevent further aggression. An emergency order issued by the Joint Chiefs of Staff included six F-102s from the 509th FIS, which touched down at DaNang AB on 5 August, and six from the Naha-based 16th FIS, which arrived simultaneously at Tan Son Nhut AB. Movement of the F-102 squadrons, in conjunction with B-57 and F-100 units, pointed out that air power would play a major part in a very long war.

No longer recognizable as an F-102A, this Deuce was one of three destroyed during a ground attack at DaNang AB during July 1965. (Fred Reiling)

Still in its gray finish, an F-102A of the 509th FIS stands alert duty at Bangkok, Thailand, in December 1965. (Joe Sutila via Jerry Geer)

A 509th Deuce shortly after landing at Bien Hoa AB, South Vietnam, during November 1967. (D. Remington via Jerry Geer)

A total of 19 F-102s, representing the 64th and 509th Fighter Interceptor Squadrons, at Clark AB, Philippine Islands, in the spring of 1969. (Larry Davis)

As expected, no enemy air attacks were flown against South Vietnam, so the reduction of F-102 units resumed. Eventually, only the 509th FIS operated the Deuce in the PACAF. By the end of June 1965, the squadron boasted an extraordinarily large compliment of F-102s to compensate and keep pace with its growing alert responsibility, not only in South Vietnam and Thailand, but throughout the Pacific. Delta Dagger deployments, which were code named "Candy Machine," were conducted on a steady rotational basis, usually for 90-day periods. Besides F-102 alert detachments at DaNang and Tan Son Nhut, others were stationed at Bien Hoa AB in South Vietnam and Udorn Royal Thai Air Force Base and Don Muang Airport in Thailand for limited air defense.

Eventually, all of the 509th's F-102s were equipped with infrared modifications. Ironically, they were never used for air defense, but did serve in a wide variety of roles. Rather than have F-102s and their pilots stand endless hours of alert duty

at DaNang AB, they were pressed into service as top cover for the 6252nd Tactical Fighter Wing, after it was activated at DaNang on 8 July 1965. The mission of DaNang's Deuces became one of providing protection against any air threat that might be mounted against the unit's multi-faceted strike and support forces. The DaNang F-102 detachment, along with TDY F-104s, was relieved by an F-4 Phantom unit in October 1965.

The first operational loss of an F-102 in Southeast Asia occurred on 24 November 1964, when a 509th Deuce suffered engine failure and crashed in South Vietnam. The first combat losses occurred when a North Vietnamese Army (NVA) "Sapper" force attacked DaNang AB during the early hours of 1 July 1965. During the assault, a demolition squad first destroyed two C-130 transports, then fired their weapons into the exhaust sections of three F-102s. Since the Deuces were parked "cocked" with their canopies open, the

This F-102A of the 509th FIS served as the mount for the commander of the 405th Fighter Wing. Seen at Udorn RTAFB in 1969, the Deuce carries an Irish theme in light of the commander's heritage. The commander's stripes were dark green, as were the shamrock and pilot's title, "Col. E.P. McNeff," painted on the nose landing gear door. (Larry Davis Collection)

This makeshift weather shelter at DaNang AB could accommodate three F-102s. (Tom Hansen)

This view of two 509th Deuces over the Southeast Asian countryside illustrates the effectiveness of the tri-tone camouflage. (USAF)

infiltrators were able to toss grenades into the open cockpits. Rockets aboard some of the burning F-102s exploded or launched towards the flight line. Besides Air Policemen, F-102 pilots and maintenance personnel, armed with hand guns and unauthorized M-16 rifles, engaged the enemy force. When it was over, one Air Policeman lay dead and several F-102 personnel were wounded. One F-102 was destroyed outright, two sustained extensive damage, and four received damage that was repairable. In addition, two C-130s were destroyed and one badly damaged.

In what was perhaps the most unusual role for the Deuce, and more an experiment than tactical necessity, Deuces from the alert detachment at Tan Son Nhut AB became involved in air-to-ground operations. The unique assignment began during mid 1965 under the code name "Project Stove Pipe." The F-102 was first used at night against the vexation known as the "Ho Chi Minh Trail," the enemies' major supply route into the south. Using their infrared seekers to lock onto heat sources, Deuces unleashed Falcon missiles against unseen targets. Radar missiles were also fired if a target presented itself. Pilots were seldom aware that they actually destroyed targets, although secondary explosions were periodically observed. Considering the number and variety of systems the Air Force pitted against the Trail, it would have been more unusual had the F-102's IR seeker not been tried.

In an even more radical deviation from their intended use as an interceptor, the F-102s switched to a day fighter role. Using the optical sight, FFAR rockets were fired at ground targets. More than 600 day sorties were flown throughout the remainder of 1965. One F-102 was shot down by ground fire during a rocket run at a ground target on 15 December. Having demonstrated its capabilities as a tactical fighter, the F-102 was periodically called upon for close support missions, especially in emergencies, since they were considerably faster than F-4 Phantoms. Although the F-102 was completely out of its element in the close support role, it did prove, during

This uncamouflaged F-102A, taxiing at Tan Son Nhut AB, South Vietnam, during 1965 had a "B" added to its buzz code, indicating its assignment to "B Flight" of the 509th FIS. A PACAF emblem was worn on the tail fin. Shortly after this photo was taken, the aircraft was shot down by ground fire. (USAF)

While assigned to the 64th FIS, this Deuce was downed by ground fire during a Combat Air Patrol mission over South Vietnam. The pilot, Capt. A.J. Sever, ejected and was rescued. (USAF)

A Deuce of the 509th's Detachment 6 taxis at Don Muang Airport, Thailand, during 1967. (Neal Schneider)

that strange period in its history, that it was capable of air-to-ground operations.

By the end of 1966, the number of aircraft committed to air defense in Southeast Asia had peaked. A total of 12 F-102s stood alert in South Vietnam (6 at Bien Hoa and 6 at DaNang) and another 12 in Thailand (6 at Udorn RTAFB and 6 at Don Muang Airport). Two of the F-102s at Don Muang stood alert specifically to guard the King of Thailand. Although highly unlikely, had the need arisen to augment the F-102 force, more than a dozen F-104s were assigned air defense as a secondary mission.

In a role for which they were marginally suited, F-102s also flew Combat Air Patrol escort missions for B-52 bombers flying against targets in North Vietnam. It was on one such mission, on 3 February 1968, that two 509th Deuces were attacked by a pair of North Vietnamese MiG-21s. Not a dogfighter, one of the F-102s, call sign "Jersey White," was shot down by an "Atoll" air-to-air missile fired by a MiG. The Deuce pilot, 1LT W.L. Wiggins, is listed as MIA.

Among the wide range of duties assigned the F-102s was escort of VIP and civilian flights, usually to "show the flag." During late 1965, one TF-102A was used as a photo chase aircraft for B-52 "Are Light" strikes against Viet Cong strongholds in South Vietnam.

Although the 509th FIS had nearly 40 F-102s in their inventory by the end of 1965, the unit found it increasingly difficult to maintain the 24 aircraft based in Southeast Asia. Relief came in the form of ADC F-102 units transferred to the PACAF. Although nearly 90 ADC F-102s had been modified during late 1965 for in-flight refueling to facilitate the long-distance deployment, the decision was made to transfer 52 aircraft to PACAF bases with the 64th and 82nd Fighter Interceptor Squadrons. Both units completed the move from U.S. bases to those in the Pacific by early June 1966. The 13th Air Force then drew upon the 64th and 509th squadrons to maintain alert detachments in South Vietnam and Thailand. The 64th replaced the 509th FIS alert detachment at DaNang on 17

June, and from there deployed Deuces to Thailand's Don Muang Airport. The parent command of the 64th and 509th FIS was the 405th Fighter Interceptor Wing, which not only managed Tactical Fighter Squadrons, but maintained five TF-102As among its Deuce fleet.

The 82nd FIS, which was under the 51st FIW, was committed primarily to PACAF assignments outside the war zone, although some combat missions were flown. Clark Air Base had become a major staging area, not only for F-102 operations, but for other aircraft types, as well. Under that concept, had it become necessary to commit additional F-102s to the war, they were already poised on the war zone perimeter. A minimum of 14 Deuces were kept on 5-minute alert, with the remainder on 1-hour call at Southeast Asia bases throughout 1967 and '68.

F-102 losses in Southeast Asia were considered minimal and proportional to the type's overall safety record. In nearly a decade of flying a wide variety of missions in the war zone, only 15 F-102s were lost. All, with the exception of one flown by the 64th FIS, belonged to the 509th, testimony to its lengthy involvement in the war. A total of 8 were operational losses, with the remainder attributed to combat. Of those, 4 were destroyed during ground attacks, 2 were downed by ground fire, and one was shot down by a MiG-21.

The see-saw deployment of the F-102 alert detachment at Bien Hoa AB ended on 25 September 1968, leaving the 509th with its detachment at Udorn. The squadron also flew a variety of missions on a sporadic basis throughout South Vietnam. The 64th FIS discontinued alert rotations in Southeast Asia on 15 November 1969 and was inactivated at Clark AB on 15 December. It then became necessary for the 509th to fill the void at Thailand's Don Muang Airport with several F-102s. The 509th ceased Deuce alert operations in South Vietnam during early 1970, however, it continued to operate the contingent at Don Muang until late May. On 24 July 1970, the squadron was inactivated. As Deuces wound down operations in Southeast Asia, their activities were usually taken over by F-4s.

A Deuce of the 82nd FIS at Naha AB on 15 January 1968. (Charles B. Mayer)

The war in Southeast Asia brought to a close the era of colorful markings and distinctive insignia familiar to USAF aircraft. In a move to thwart communist agents from tracking Air Force units on alert, an order issued during the summer of 1964 directed that all unit and colorful markings be removed from PACAF interceptor aircraft. Initially, only a small PACAF emblem appeared on both sides of the vertical fin. When a large number of 64th and 82nd FIS F-102s were modified with mid-air refueling equipment during late 1965 for their impending transfer to PACAF bases, they also received a camouflage paint scheme, which eventually became standard on aircraft operating in Southeast Asia. The "tri-tone" scheme was introduced in the Southeast Asian theater during early 1966, and all 509th FIS F-102s had donned the subdued livery by year's end. Those who admired how the "interceptor gray" finish complimented the Delta Dagger's sleek lines were undoubtedly dismayed by the application of camouflage. Some F-102s of the 82nd FIS arrived in the Pacific theater during February 1966 wearing black-stenciled ADC and squadron emblems on their freshly-camouflaged tails. In keeping with Air Force policy of the period, tail codes were assigned to the F-102 units during mid 1967.

F-102 UNITS IN SOUTHEAST ASIA

UNIT	TAIL CODE	COMMAND	PACAF BASE	YEARS IN THEATER
64th FIS	PE	405th FIW	Clark AB, P.I.	1966-1969
82nd FIS	NV	51st FIW	Naha AB, Okinawa	1966-1971
509th FIS	PK	405th FIW	Clark AB, P.I.	1960-1970
16th FIS		51st FIW	Naha AB, Okinawa	1964

F-102 LOSSES IN SOUTHEAST ASIA

DATE	UNIT	A/C SERIAL	REMARKS
11-24-64	509th FIS	56-1189	engine failure over SVN
7-1-65	509th FIS	55-3371	ground attack at DaNang
7-1-65	509th FIS	55-1161	ground attack at DaNang
7-1-65	509th FIS	56-1182	ground attack at DaNang
12-15-65	509th FIS	55-3373	ground fire over SVN
8-19-66	509th FIS	56-1110	operational
12-9-66	509th FIS	56-1162	operational
12-14-66	64th FIS	56-1389	ground fire over SVN
1-15-67	509th FIS	55-4036(TF)	ground accident Thailand
4-2-67	509th FIS	55-3362	engine failure over SVN
5-12-67	509th FIS	55-1165	ground attack at Bien Hoa
2-3-68	509th FIS	55-1166	shot down by MiG-21
7-17-68	509th FIS	56-0963	engine failure over SVN
9-16-68	509th FIS	56-0970	collision with F-4 Thailand
1-8-69	509th FIS	56-1186	operational over Thailand

As intended, the application of warpaint greatly disrupted the sleek, classic lines of the F-102 as evidenced by this trio flying over Southeast Asia. (USAF)

F-102s of the Air National Guard

In view of Air Force policy to maintain a high degree of proficiency in the Air National Guard, its aircraft inventory was continually upgraded to keep it on a par with the Air Defense Command. With USAF Headquarters' blessing, ADC and ANG leaders drew up plans in 1958 for the F-102's debut in the Air National Guard, which was to occur in late 1961 with the conversion of four units. The aircraft would come from ADC squadrons scheduled to trade in their F-102s for more modern types. It was further anticipated that foreseeable deactivations of ADC fighter interceptor squadrons would yield additional F-102s for Guard use.

At odds with that plan was a 1959 NORAD proposal for the establishment of 19 ANG alert squadrons. They were to be permanently positioned along U.S. northern and southern borders, plus other strategic areas, to fill the gap in interceptor coverage. NORAD planners mistakenly assumed that there were sufficient F-102s to equip such units. Although ADC squadrons were scheduled for upgrades, primarily to F-101B and F-106A aircraft, thereby freeing F-102s for ANG assignment, the solution for implementation of the NORAD scheme came from reducing the number of ADC fighter interceptor squadrons.

The deactivation of units began during July 1959, and within one year, the number of ADC fighter interceptor squadrons dropped from 27 to 15. The changeover resulted not only in more F-102s available, but gave the Minutemen a much larger role in North America's air defense. That expanded responsibility actually took shape during the early 1950s, when ANG mobilization assignments were shifted from the Tactical Air Command to the Air Defense Command. Its part in the alert program and sweeping conversion to jets ensured the Guard a place at the forefront in the air defense system. The Air Defense Augmentation Program, a result of Cold War apprehension, went into effect on 15 August 1954, and by year's end, a total of 17 squadrons were at alert status.

By July 1958, all of the Air National Guard's 84 combat squadrons were jet-equipped. The F-102 was one of four new aircraft types to join the Guard during 1960. Its introduction came on the heels of budget cutbacks, which brought about a gaining command concept adopted in July of that year. The terms of the proviso had ANG units placed under the direct control of air commands that would gain them in the event they were mobilized. Thus, the ADC became the parent command of Guard fighter interceptor units. During 1961, when the Guard's fronts were further expanded, the ANG boasted 32 fighter interceptor squadrons, 23 of which would operate the F-102. Owing to their vastness, the states of Texas and California each hosted two squadrons. Although the Air Force initially proposed to equip four ANG squadrons by the end of 1960, ADC accelerated the process to alleviate the number of surplus F-102s at its bases.

First to take on the Deuce was the 182nd Fighter Interceptor Squadron of the Texas Air National Guard, its first F-102 having arrived at Kelly AFB on 4 June 1960. Following closely was its sister unit, the 111th FIS at Ellington AFB, which took delivery of its first F-102 in August. Louisiana's 122nd FIS and Florida's 159th began trading in their F-86Ls for F-102s in July. By the end of 1960, the Air National Guard of Pennsylvania, South Dakota, and Hawaii had also converted to the more capable F-102.

The markings of this F-102A were an exception to those worn by the Minnesota Air Guard's 179th FIS. The unit became known as "The Aces of Deuces" when it took top honors at the 1970 William Tell competition, the only meet in which it participated. (Maj. Gen. Wayne C. Gatlin)

A Deuce of the Wisconsin Air Guard's 176th FIS in September 1968 (WI ANG)

A number of silver-painted F-102s found their way into Air National Guard squadrons, such as this example of Idaho's 190th FIS, seen here at Edwards AFB in October 1962. (Baldur Sveinsson Collection)

Montana's first F-102 arrived at Great Falls on 30 July 1966, ending an 11-year F-89 era. Following Air Guard service, this immaculate example went to a private college as an instructional airframe. (Alec Fushi via Terry Love)

A total of 21 ANG fighter interceptor squadrons were operational with the F-102 by 1966, with the exception of Maine and New York, which received theirs in 1969 and 1972, respectively. The New York Guard's term with the Deuce was a hiatus from Military Airlift Command control. Most units had exchanged aging F-89J Scorpions or F-86L Sabres for the high performance Deuce. The reverse occurred in Tennessee and South Carolina units where F-102s replaced F-104 Starfighters the Air Force needed to equip newly formed wings. A squadron typically comprised 25 aircraft, two of which were TF-102As. Much like what occurred in the ADC, large numbers of aircraft were often rotated through ANG units as models were continually upgraded. One TF-102A (s/n 56-2353) served the ANG in five states before ending its career as a display.

When communists began erecting the Berlin Wall in 1961, the U.S. response included the deployment of eight TAC squadrons to Europe, beginning in early September. President Kennedy then ordered 28 ANG squadrons to active duty as long-term replacements for the TAC units. Under Operation STAIR STEP, 11 of those squadrons were deployed to USAFE, three of which were fighter interceptor squadrons. South Carolina's 157th FIS was reassigned to the 65th AD on 25 November 1961 for duty at Moron AB, Spain, while Tennessee's 151st FIS and Arizona's 197th FIS were reassigned the same date to the 86th AD for operations from Ramstein AB, Germany. While based in Europe, the ANG squadrons conducted regular operations with USAFE F-102 units. Of special interest was the coincidental approval to modify F-102s of the first six ANG squadrons to accept the GAR-11 nuclear missile.

The only other ANG deployment beyond the continental U.S. during the troubled period was to Alaska for training. Air National Guard units were not activated during the Cuban

This F-102A of Idaho's 190th FIS wears the Air Force Outstanding Unit Award on its tail fin. The 190th was the first Guard unit to win the award consecutively—in 1970 and '71. (Ken Buchanan)

Nose art applied to F-102A serial number 55-3360 of the 199th FIS. (Jack M. Friell)

A silver-painted Deuce of the 122nd FIS, 159th FIG at Langley AFB during June 1970. (Don Larsen via Terry Love)

Different tail fin tip colors of these 199th FIS F-102s signify their assignment to flights within the squadron. (Jack M. Friell)

missile Crisis during fall 1962, although they were placed on increased alert status. Some did serve alongside ADC F-102 units that were tactically dispersed during the tense period.

The Air National Guard Air Defense Alert Program provided a vital boost in ADC capability throughout the 1960s. Fighter interceptor squadrons provided around-the-clock alert coverage with two or three aircraft prepared for takeoff within five minutes. Frequent air defense exercises, which included local training sorties, scrambles, and Operational Readiness Inspections (which, in themselves, were grueling, realistic combat scenarios) became a way of life for ANG Deuce units.

Of great importance in maintaining a state of readiness was the Guard's participation in the Worldwide Weapons Meet, better known as William Tell. Involvement in the competition seemed a good indicator of the Guard's increased capabilities. Air National Guard F-102 teams participated in six meets beginning in 1961 with the 182nd FIS from Texas. During William Tell 1963, Pennsylvania's 146th FIS, the only

ANG entry, captured first place in the F-102 category. South Carolina's 157th FIS was the lone ANG competitor in the 1965 meet, however, that would change dramatically for the 1970 William Tell, in which all F-102 entries hailed from ANG units. First place went to Minnesota's 179th FIS, which also won the aircrew maintenance and weapons controller awards. In the 1972 meet, Wisconsin's 176th placed not only first in the F-102 category, but won the Top Aircrew Award. The 1974 William Tell was the last in which F-102s competed as shooters. It was also the first in which ANG units won top honors in all three categories. The 190th FIS from Idaho, the only Deuce unit to take part in two meets, placed first in the event.

Unique to Wisconsin's 115th Fighter Group was the "Deuces Wild" precision flying team, composed of 176th FIS pilots and ground crews. Conceived in 1969, the all-volunteer team exemplified the skills practiced daily by citizen-soldier interceptor crews as they took on more responsibility in air defense.

F-102As of the 179th FIS fly a tight echelon formation. The unit's first Deuce arrived on 22 November 1966 from Goose Bay, Labrador, wearing the orange markings, which became the standard throughout the 179th's Deuce era. (Maj. Gen. Wayne C. Gatlin)

Trimmed with blue and white tail bands, wing tips, and wing fences, a trio of Montana Air Guard Deuces traverses a U.S. mountain range. The aircraft in the foreground wears a double blue band, edged with white, while the Deuce in the background displays a single fuselage band, indicating a squadron staff aircraft. (Author)

Nose art details of F-102A serial number 55-3366 assigned to Hawaii's 199th FIS. (Jack M. Friell)

This TF-102A was one of 8 flown by the 196th FIS of the California Air Guard throughout the time it operated Deuces. Aircraft assigned to the unit were commonly interchanged with those of its sister unit, the 194th FIS. (Hugh Muir via Terry Love)

At the beginning of 1969, 374 of the Guard's 414 interceptors were F-102s. By the end of the decade, F-102s in ANG squadrons were replaced in favor of fully upgraded models, many of which were drawn from surplus stocks in Southeast Asia and Europe. It was also common for aircraft to be pulled from squadrons in groups for upgrades and replaced with improved models from air material areas.

Despite the readiness displayed by ANG units, an incident in 1971 prompted a quick reevaluation of the nation's air defense network, if only on a temporary basis. On 26 October, a control tower along the southern U.S. coast received a radio message from "Cubana 877" requesting landing instructions—the first indication that the Cuban aircraft was only 25 miles from New Orleans. Somehow, the AN-24 had slipped through the air defenses undetected and, as expected, the press had a field day with it. "Where was our southern air defense," they asked. The rapid response included three F-102 alert detachments to bolster southern air defenses. A detachment from California's 194th FIS was positioned at

Davis-Monthan AFB Arizona, another formed off of the 111th FIS at Ellington AFB, Texas, while a third, from Florida's 159th FIS, set up shop at Naval Air Station New Orleans. A pair of F-106s, which was placed on 5-minute alert at Florida's Tyndall AFB, was added for good measure, however, ANG alert crews, confident in their ability to handle the matter, felt the Delta Darts were "too fast for Cuban targets."

Although Texas' 111th FIS flew the F-102, it retained its T-33As for the ANG Jet Instrument School. During January 1970, the unit took on the added responsibility of training F-102 crews, forming the ANG's F-102 Combat Crew Training Squadron. In May 1971, F-101B/Fs were added to the training program, replacing the F-102s of the 111th.

Transitions in ANG fighter interceptor squadrons to newer aircraft resulted not only in extensive reorganizations, but often meant a change in roles, as well. While three units replaced the Deuce with its modernized stablemate, the F-106, others switched from air defense to tactical roles with Phantoms and Corsair IIs. California's 196th FIS and Wisconsin's 176th FIS

An F-102A of California's 196th FIS in 1968. (Nick Williams)

Markings that appeared on Deuces of the 194th FIS in 1968 included an Air Force Outstanding Unit Award. (Nick Williams)

Hawaii's 199th FIS used three styles of tail fin markings for its F-102s. This distinctive pattern first appeared in mid-1968 and remained until the unit switched to F-4s. (Terry Love)

After service with California's 196th FIS, this F-102A went on to serve as a full-scale target drone. (D. Slowiak via Marty Isham)

The Connecticut Air Guard exploited the F-102's large tail surface by applying a dramatic, stylized bird caricature. Billing themselves as "The Flying Yankees," the 118th FIS displayed the title on the aircraft's drop tanks. (Larry Davis Collection)

An F-102A of Florida's 159th FIS in September 1966. The unit followed the common practice of placing the last three numbers of the aircraft's serial number on the interior face of the speed boards. (Richard Sullivan via Stephen Miller)

When markings were reduced with the introduction of camouflage, the lightning bolt usually worn on the tail band of Florida's Deuces was reapplied to the fin's tip. (L.B. Sides)

This well worn "Tub" of the 159th FIS did not display its serial number, which was 56-2329. (Hugh Muir via Terry Love)

A TF-102A of the Hawaii Air National Guard in 1968. (Nick Williams)

Like a number of F-102s flown by the 199th FIS, this Deuce wore nose art. (Nick Williams)

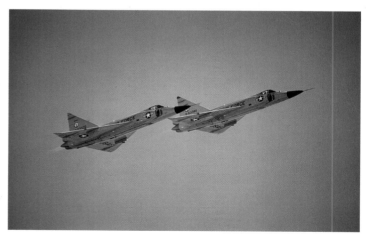

This pair of Deuces represent two different flights within Hawaii's 199th FIS as indicated by their tail fin tip colors. The squadron provided the sole air defense coverage of the Hawaiian islands. (Nick Williams)

A well-maintained F-102A of South Dakota's 175th FIS. (Charles B. Mayer)

South Carolina Air Guard Deuces had small orange squares applied to their wing fences. The unusual markings were carried over to the camouflage scheme. (Hugh Muir via Terry Love)

Beginning in December 1969, South Carolina's 157th FIS exchanged 21 of its F-102s for upgraded models, most of which appear in this view. (Larry Davis Collection)

A moon and palmetto tree, symbols incorporated into the South Carolina state flag, were added to the tail markings of this F-102A, seen here in July 1973. (Norm Taylor)

This Deuce of the 123rd FIS, seen here at Portland Airport in July 1968, broke from the norm in having its commander's stripes painted green versus red. (Peter Lewis via Marty Isham)

Though this F-102A wears the Pennsylvania Air Guard scheme, the ace of spades emblem appears on the forward fuselage, announcing its previous use with the Washington ANG. (L.B. Sides)

An F-102A in the original markings used by the 146th FIS. (L.B. Sides)

Having newly arrived to the 146th FIS, this Deuce had temporary identification applied over markings from its previous assignment, Minnesota's 179th FIS. (Candid Aero-Files)

Snow-covered mountains make an impressive backdrop for this Deuce of North Dakota's 178th FIS, also known as "The Happy Hooligans." (ND ANG)

As aircraft were passed between squadrons, markings from previous units were often retained until new ones were applied. This Pennsylvania Air Guard Deuce still wears the chevron familiar to the 61st FIS. (Larry Davis Collection)

The striking marking scheme applied to Pennsylvania Air Guard Deuces enhanced even the Tub's broad profile. The silver-tipped black drop tanks bore the title "112th FTR GP" and an Air Force Outstanding Unit Award on a white stripe. (Hugh Muir via Terry Love)

A TF-102A of the 186th FIS in 1971. (Author)

Compared to other ANG fighter interceptor squadrons, New York's 102nd FIS was late in acquiring the F-102. It wasn't until fall 1972 that the unit exchanged its KC-97L tankers for the Deuce. (Jack M. Friell)

A Wisconsin Air Guard Deuce is prepared for a mission at Truax Field. The aircraft's Ram Air Turbine is visible below the fuselage. (WI ANG)

The 182nd FIS of the Texas Air Guard flew F-102s throughout the 1960s from Kelly AFB. Nearby was the historic Alamo, which was used in the squadron's emblem, seen painted on this aircraft's drop tank in September 1967. (Stephen Miller)

The Texas 182nd FIS was the first ANG unit to receive the F-102 in a sweeping program that placed more reliance on Guard involvement in America's air defense. (Richard Sullivan via Stephen Miller)

A TF-102A of the 111th FIS at Ellington AFB, Texas. (John Guillen)

Its pilot seat removed, a 134th FIS F-102A is serviced at Burlington Airport, Vermont. (Lionel Paul)

When EB-57B/Es replaced Vermont's Deuces, the 134th FIS assumed the role of defense system evaluation. The Vermont Air Guard called themselves "The Green Mountain Boys" and proudly advertised the nickname on their aircraft. (L.B. Sides)

Having been fully upgraded with all the modifications specified for F-102s, this trainer stands alert at the 178th's home base. (Lionel Paul Collection)

An F-102A of the 175th FIS on display at Richards-Gebauer AFB in September 1967. (AAHS From Clyde Gerdes)

Like many active duty and ANG units, Vermont's 134th FIS used colored tail fin tips to identify flights within the squadron. In addition, this Deuce wore a flashy design in the flight color on the drop tank during 1967. (Nick Williams Collection)

The state of Oregon's Air Guard operated Deuces from 1965 to 1971. This Tub is seen at the 123rd's home base, Portland Airport, during June 1970. (AAHS Collection)

A pair of 182nd FIS Delta Daggers are "cocked" in open-ended alert barns at Kelly AFB, Texas. (TX ANG)

The double meaning of the Hawaiian word "Aloha" had its full effect on 22 October 1976, when a pair of Delta Daggers of Hawaii's 199th FIS flew their last operational mission in the company of their replacements, a pair of F-4C Phantoms. Flying the lead in this view was Col. James R. Ashford, commander of the Hawaii Air Guard's 154th Tactical Fighter Group. The F-102 remained at the unit's home base, Hickam AFB, as a permanent display. (HI ANG)

A Deuce of the 116th FIS in May 1967. (Robert Burgess via Stephen Miller)

The commander of the Texas 182nd FIS, Col. Charles A. Quist, Jr., poses with one of the squadron's F-102As. (TX ANG)

A Christmas tree was added to the blue and white tail markings of this 186th FIS Deuce in 1966. (MT ANG)

Its missiles extended on their racks for display, this Deuce of the 122nd FIS is seen during May 1965. (Lionel Paul)

The legend "First Class" on the external fuel tank of this 190th FIS Deuce was in keeping with the unit motto, "First Class, Or Not At All." (L.B. Sides)

An F-102A of Washington's 116th FIS undergoes a tire change. Deuces of the 116th wore their unit emblem, an ace of spades playing card pierced by a dagger, on the forward fuselage. (Lionel Paul)

Having inherited most of its aircraft from Alaska's 317th FIS when that unit was inactivated, the 176th FIS of the Wisconsin Air Guard retained the high-visibility markings specified for aircraft that operated in snowy regions. Wisconsin Deuces wore special markings on their drop tanks proclaiming their skills at the 1972 William Tell competition. (Baldur Sveinsson)

Details of the 146th FIS black and white tail fin markings. (Lionel Paul Collection)

Displaying a slight variance in markings, F-102s of the 146th FIS are lined up at Greater Pittsburgh Airport. (Robert F. Dorr Collection)

Any photo depicting F-102s assigned to the Maine Air Guard is rare, considering that the state's 132nd FIS operated them for only five months. The majority of Maine's F-102s were acquired from the Holland-based 32nd FIS when that unit transitioned to the F-4E. (ME ANG)

This Deuce of the 146th FIS served as a billboard for multiple unit awards. Displayed on the tail fin are ADC and Air Force Outstanding Unit Awards, while a first place William Tell emblem and four "Firebee" drone silhouettes are featured on the nose. In addition, the pilot's and crew chief's titles appear on "keystones" superimposed over a red triangle on the nose landing gear door. (Robert F. Dorr Collection)

A 176th FIS Deuce in the distinctive markings applied by the 317th FIS for contrast against Alaska's terrain. (R.M. Hill via David Hansen)

Fifteen F-102As of Hawaii's 199th FIS on the flight line during the summer of 1961 in the squadron's original markings. (Marty Isham Collection)

Like their active duty counterparts, Air Guard Deuces suffered a high incidence of nose gear failure. This F-102A of South Dakota's 175th FIS lands on a foamed runway after the nose gear failed to extend during summer 1969. Damage was limited to the radome, and the Deuce was quickly returned to service. (SD ANG)

An F-102A of the 178th FIS undergoes maintenance at North Dakota's Hector Field during the late 1960s. (ND ANG)

An F-102A of Tennessee's 151st FIS at Knoxville during July 1964. Having traded in their F-104s for F-102s, the 151st flew the Deuce for only 16 months beginning in March 1963. (Marty Isham Collection)

underwent a more dramatic change when they exchanged their F-102s for prop-driven 0-2A light observation aircraft. California's switch to a less aggressive type in 1975 reportedly was based on anti-war sentiment and subsequent political pressure. Other radical changes included Tennessee's conversion to KC-97L tankers and New York's to rescue versions of the C-130 Hercules and H-3 helicopter.

Maine's interim use of the F-102 was the type's shortest-lived assignment in the Guard. The first Delta Dagger for the 132nd FIS arrived on 4 July 1969 to replace the unit's F-89J Scorpions, only to be replaced during mid-November by the F-101B Voodoo. At the time that three active duty F-101 squadrons were slated for inactivation, Maine had been selected to receive the Voodoo, and the withdrawal date for F-89s was firm. After the brief stay, the F-102s were passed to Florida's 159th FIS. In stark contrast, the award for longevity goes to Hawaii's 199th FIS, which operated the Deuce for nearly 17 years. Deuces flew alongside the unit's F-4Cs as late as 1976 before they were retired.

Deuces typically operated in pairs, including takeoffs and landings, a practice that narrowed the type's safety margin. This 157th FIS duo flies over South Carolina in June 1970. (SC ANG)

Underscoring the high risk in fighter interceptor operations, a total of 44 TF/F-102As were lost during service with the Minutemen. Hardest hit was Florida's 159th FIS with nine Delta Daggers destroyed, including the Guard's only two TF-102A losses. Maintaining a state of readiness had an even higher price—the loss of 14 Air National Guard F-102 pilots.

As with the Air Force, the F-102 had the distinction of serving not only as the Guard's first delta-wing, supersonic interceptor, but as the backbone of ANG air defense.

AIR NATIONAL GUARD F-102 UNITS

UNIT	STATE	F-102 TIME FRAME	LOCATION
102 FIS/106 FIG	New York	5 May 1972-31 Dec 1975	Suffolk County
111 FIS/147 FIG	Texas	Aug 1960-17 Sept 1974	Ellington AFB
116 FIS/141 FIG	Washington	29 June 1965-24 Nov 1969	Geiger Field
118 FIS/103 FIG	Connecticut	2 Nov 1965-27 July 1971	Bradley Field
122 FIS/159 FIG	Louisiana	July 1960-8 Dec 1960	NAS New Orleans
123 FIS/142 FIG	Oregon	16 Nov 1965-16 Apr 1971	Portland Airport
132 FIS/101 FIG	Maine	9 Apr 1969-14 Nov 1969	Bangor Airport
134 FIS/158 FIG	Vermont	13 Aug 1965-6 May 1974	Burlington Airport
146 FIS/112 FIG	Pennsylvania	Nov 1960-31 Dec 1975	Greater Pittsburgh Airport
151 FIS/134 FIG	Tennessee	4 Mar 1963-21 July 1964	McGhee-Tyson Airport
152 FIS/162 FIG	Arizona	23 Feb 1966-28 May 1969	Tucson Airport
157 FIS/169 FIG	South Carolina	June 1963-23 Mar 1975	McEntire ANGB
159 FIS/125 FIG	Florida	July 1960-29 Aug 1974	Cole Imeson Airport
175 FIS/114 FIG	South Dakota	Oct 1960-1 July 1970	Sioux Falls Airport
176 FIS/115 FIG	Wisconsin	7 Feb 1966-11 Oct 1974	Truax Field
178 FIS/119 FIG	North Dakota	29 June 1966-19 Nov 1969	Hector Field
179 FIS/148 FIG	Minnesota	22 Nov 1966-22 Apr 1971	Duluth Airport
182 FIS/149 FIG	Texas	4 June 1960-31 July 1969	Kelly AFB
186 FIS/120 FIG	Montana	9 June 1966-9 June 1972	Great Falls Airport
190 FIS/124 FIG	Idaho	April 1964-31 Dec 1975	Boise Air Terminal
194 FIS/144 FIG	California	July 1964-28 June 1974	Fresno Air Terminal
196 FIS/163 FIG	California	15 Apr 1965-31 Dec 1975	Ontario Airport
199 FIS/154 FIG	Hawaii	5 Dec 1960-28 Jan 1977	Hickam AFB

A pair of early model F-102As of the 190th FIS over Idaho in May 1968. (ID ANG)

Deuces of the 122nd FIS. Small stripes on the speed boards of the trailing aircraft may have indicated a staff officer's aircraft. (LA ANG)

A pair of Deuces on takeoff was an impressionable sight and sound. This duo departs Tucson Airport during the late 1960s. (Robert F. Dorr Collection)

This pair of F-102s wears the unmistakable markings of Florida's 159th FIS. (FL ANG)

A pair of 118th FIS F-102As flies a tight formation landing sequence. Visible on the Deuce in the background are the dual data link antennas. (Thomas S. Cuddy via Leo Kohn)

This trio of 182nd FIS F-102As wear the second style of markings used by the squadron. The 182nd's sister unit, the 111th FIS, used blue chevrons. (Marty Isham Collection)

Emblazoned with yellow markings, three F-102As and a TF-102A of the 152nd FIS fly over Arizona. (Marty Isham Collection)

During an Alaskan deployment, Mount Denali of the McKinley range serves as a backdrop for a flight of four from the 178th FIS. Different colored speed boards indicate aircraft from two flights. (USAF)

Ground support equipment was towed through deep snow at Bradley Airport, Connecticut, to service these F-102As of the 118th FIS. (USAF)

A number of requests for flyovers prompted members of Wisconsin's 176th FIS to form a flight demonstration team, first with F-89s and later F-102s. Called the "Deuces Wild," the team used four aces and a deuce as a logo and flew with five aircraft, numbered "01" through "05." After several appearances, Air Force officials put an end to the team with the pronouncement that the "Thunderbirds" would be the only formation team and all others would disband. To commemorate the team, this Deuce wore red and white checked speed boards and a pair of deuces playing cards superimposed over a broad dark-colored band, in addition to the unit markings. Like the team, the markings were short-lived, and the aircraft was placed on permanent display. (WI ANG)

Texas Governor George W. Bush, a former Deuce pilot, prepares for a mission during the time he flew Texas Air National Guard F-102s. (George Bush Presidential Library)

Special Projects

Besides the battery of tests undergone by the initial batch of F-102s, the type became involved in a wide range of tests and special programs throughout its service life. As a result, the F-102 not only made its mark in aviation history, but was instrumental in the development of numerous systems for generations of future aircraft. As a dramatic new concept in interceptor design, the F-102 attracted the attention of a variety of agencies that were eager to expand its use with the application of unique systems. As a high performance aircraft, the Deuce was a natural for assuming the role of high speed and tactical research platform. A look at the more unusual aspects of the F-102's history reveals a substantial number of aircraft, some oddly transformed, whose purpose deviated dramatically from that of air defense.

One interesting spinoff of the F-102 stemmed from a proposal by Convair in 1956 for a tactical strike version known as the F-102C, and briefly, the F-102X. By installing an advanced J-57-P-47 engine, which would boost the top speed to Mach 1.3 and increase the ceiling by 3,000 feet, Convair expected the new model to fill a possible gap between the end of the F-102's usefulness and the introduction of the F-106. Structural modifications, improved fire control systems, and a weapons load of four Falcon missiles and one MB-1 Genie were to guarantee its success in the tactical arena. Two F-102As were modified as YF-102C engineering test beds, however, the concept was rejected by the Air Force in April 1957.

Shortly after the Deuce became operational, the Air Force wasted little time in committing it to test programs. As a result of the Air Force's pledge to support the Atomic Energy Com-

mission, F-102s became the first supersonic aircraft to participate in the testing of nuclear devices. The initial plan called for six F-102s of the 327th FIS to fly through the radioactive cloud of a nuclear detonation at the Nevada Test Site as part of the 1957 tests, called "Operation Plumbbob." Since Air Force interceptor pilots were expected to penetrate nuclear clouds while attacking enemy aircraft, the flights were intended to indoctrinate pilots and discover the effects of radiation on F-102s. After extensive research, scientists of the Air Force Special Weapons Center at Kirtland AFB, New Mexico, assured the pilots that they were in no danger from fallout.

Six F-102s flew from the 327th's home base, George AFB, to Nellis AFB, Nevada, in June to take part in Plumbbob. Since the unit was the first to convert to the F-102, aircraft and parts were in short supply and getting the squadron combat ready had proven challenging. In view of the arduous task and numerous weather delays, three F-102s were withdrawn from the test to keep the squadron on task with crew and maintenance training. That left three Deuces to fly through the cloud.

When the 20-kiloton device was finally scheduled for detonation, one of the F-102s experienced engine start problems. Since precise timing of the entire flight was critical, the Deuce was pulled off of the mission. After observing the detonation and build-up, the two remaining pilots, Maj. Budd H. Butcher and Lt. Richard N. Satterfield, flew through the radioactive "mushroom" at 20,000 feet. The F-102s were so radioactive "hot" that they were never returned to the squadron.

As the premiere F-102 squadron, the 327th went on to distinguish itself with another "first" as one of three ADC squad-

Examples of the first four jets of the Century Series, all 1953-vintage aircraft, fly over the Edwards AFB range where they were based for tests in 1957. (NASA)

Perhaps Convair knew something that no one else did when it proposed the F-102C as an interim interceptor. It was to go into service when the F-102 had served its usefulness, but prior to the appearance of the F-106. Two were so modified, the other aircraft having been serial number 53-1797. Number 53-1806 is seen here at Sheppard AFB during June 1960. (Merle Olmsted via David McLaren)

This was one of three F-102s assigned to the Air Force Cambridge Air Research Center at Hanscom AFB during 1957 for Test Support. Foremost among their duties were trials of the SAGE system. Unusual was the aircraft's buzz number positioned on the nose. (Jim Burridge)

Pilots of the 327th FIS conduct a briefing prior to flying through an atomic cloud during the "Plumbbob" tests in Nevada during 1957. From left to right are Deuce pilots Avery, Morton, Satterfield, Ferguson, Speer, and Butcher. Only Satterfield and Butcher completed the flight. (USAF)

rons that participated in the 1957 Bendix Race. First held in 1931, the challenging competition attracted the best pilots in the nation. Six pilots, two each from squadrons in the three ADC commands—Eastern/323rd FIS, Central/11th FIS, Western/327th FIS—shared aviation history with notables from past events, which included Jimmy Doolittle, Roscoe Turner, Amelia Earhardt, Paul Mantz, and Laura Ingalls. The time had come for a new breed of pilots, flying supersonic Delta Daggers, to step into the spotlight to prove the mastery of their profession. The coveted Bendix Trophy worked its magic as the epitome of demands made on both pilot and airplane.

The race itself was a 615-mile speed course between Chicago and Washington, D.C. The goal of the 19th event was for the F-102 to surpass the 666 mph record set in the previous year's race by an F-100C. Flying one of four of the six F-102s with external fuel tanks, Capt. Kenneth D. Chandler of the 11th FIS bolted from Chicago's O'Hare Field on 28 July. Still on afterburner, he climbed to only 25,000 feet, av-

During cold weather tests at Ladd AFB, Alaska, a Wolverine heater is connected to this F-102A's ventilating system, while an auxiliary power unit (APU) is positioned opposite. (General Dynamics)

A landing F-102's image is captured in the mirror of a U.S. Navy landing system set up at Edwards AFB to assist with landing gear load tests during July 1958. By enabling the pilot to maintain the proper glide path during different landing configurations, the number of landings during the tests was significantly reduced. (General Dynamics)

Convair Engineering Test Laboratory engineers and Air Force technicians at the Experimental Track Branch, AFFTC, Edwards AFB ready the TF-102A combat trainer test sled for a seat-ejection demonstration. The tests compared the effectiveness of rocket propulsion versus an explosive charge to eject TF-102A pilots and seats. The large access doors were standard on TF-102As. (General Dynamics)

As an F-102A rocket sled hurtles down the track, its seat and dummy are ejected in a fiery blast. Dummies were heavily instrumented to measure wind blast, acceleration, and deceleration forces. Twin booms extending behind the seat stabilized the seat during its rocket-propelled separation from the sled platform. (General Dynamics)

eraging 679.053 mph during his supersonic sprint. Just 54 minutes, 45.5 seconds later, he flashed past the finish marker at Andrews AFB, Maryland. Less than two minutes had elapsed between Chandler's arrival and the 6th place F-102.

Another prestigious award for which the F-102 qualified was the Hughes Achievement Trophy, which was sponsored by the Hughes aircraft firm and awarded to the top Air Force squadron with the air defense/air superiority mission. The annual award has been claimed by eight F-102-equipped fighter interceptor squadrons since Hughes started the program in 1953. The recipients were as follows: 1956-317th FIS, 1958-31st FIS, 1960-460th FIS, 1963-497th FIS, 1965-317th FIS, 1966-32nd FIS, 1968-64th FIS, and 1970-57th FIS.

Deuce pilots seemed comfortable with the fact that they strapped into a lightweight Weber ejection seat, which could fire them clear of the aircraft in two seconds. Nevertheless, a keen interest in crew safety and Convair's continuous efforts to improve their product brought about an extensive egress test program. Conducted jointly with the Air Force, the elaborate program led to the fabrication of some most unusual F-102 derivatives. Joint testing of interceptor egress systems by the Convair Engineering Test Laboratory and the Air Research and Development Command got under way in 1957.

At the heart of the tests were rocket sleds specially constructed from F-102 airframes, which ran on steel rails secured to concrete footings. Propelling the sleds to incredible

Its forward half angled upward to simulate a nose-high attitude, an F-102A rocket sled sheds its canopy just milliseconds before the seat and dummy are ejected during a test run. (General Dynamics)

This view of the F-102A rocket sled shows the 16 rocket ports and aft stabilizing fin. (General Dynamics)

Convair technicians install an experimental ejection seat into the cockpit of an F-106 during interceptor egress trials. (General Dynamics)

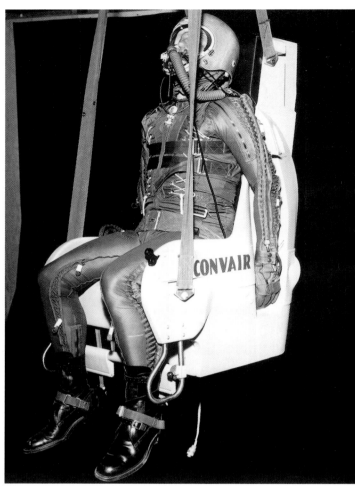

A Convair production test pilot tests a supersonic contour seat developed to ensure a safe escape from the F-102A. The pilot stayed with this seat during descent. It featured recessed cavities, wrist and foot lanyards, torso harness, helmet inertia reel, stabilizing parachute, oxygen supply, survival kit, and a rocket nozzle to blast the seat clear of the aircraft. (General Dynamics)

An F-106B was used for egress system trials, along with the F-102 rocket sleds. Pictured here during a test ejection is the first produced two-seat F-106B. A second specially modified Delta Dart, serial number 57-2516, was also used. The "B" seat shown here ejected the pilot in a supine position. (General Dynamics)

land speeds were 16 RATO-type propulsion bottles. The forward half of the sleds developed from F-102As could be angled upward to simulate a nose-high attitude. They were used primarily for supersonic testing of the twin-boom configuration of the "B" ejection seat at the ARDC's rocket sled track on Hurricane Mesa in southern Utah. The track ended near the mesa's edge, where the seat and dummy were parachuted to a canyon mesa floor 1,500 feet below. The "B" seat was developed and tested by the Industry Crew Escape System Committee for the Century Series aircraft.

Concurrent tests with a TF-102A rocket sled were conducted at the Experimental Track Branch of the Air Force Flight Test Center at Edwards AFB. The trials compared the effectiveness of rocket propulsion versus an explosive charge in catapulting seats and pilots from the side-by-side trainer. The two systems evaluated during the comparison were the standard M3 cartridge and the newly-developed RESCU Mark I rocket catapult, the latter of which attained much greater alti-

Designated a JF-102A, this Deuce shared test duties with a YF-102A at the Dryden Research Center during the late 1950s. (NASA)

This Deuce wore NACA markings in addition to those of the Air Force during its three-year stint at the Ames Research Center. (NASA)

tude. Tests were also conducted with the "B" seat, called the "Aerial Bobsled," using a miniature F-102A model at the Southern California Co-operative Wind Tunnel at Pasadena. The egress systems tests culminated with inflight ejections from a specially modified NF-106B.

Besides the NACA's (later NASA) extensive involvement with early model F-102s, the agency continued its acquisition of Deuces for various research programs. Between 1954 and 1974, a total of nine F-102s occupied the ramps at four NACA/NASA facilities to fulfill a number of special needs.

The NACA received its first Deuce, a YF-102A, in September 1954. The early variant made a total of 104 test flights at the Dryden Flight Research Center before it was passed to the adjoining Edwards AFB in 1958. During 1956, the YF-102A was joined by a JF-102A, which made 74 test flights, 48 of which focused on delta-wing performance. Test pilots Jack McKay and Neil Armstrong flew approach and abort maneuvers in the aircraft for the ill-fated X-20 "Dyna-Soar" program. Four Deuces (two F-102As and two trainers), the largest number assigned to a NASA facility, went to the

Manned Spacecraft Center at Houston, Texas, for the Astronaut Proficiency Flying Program. Based at Ellington AFB, the Deuces were flown extensively for nearly two years by the first two groups of astronauts, totaling 15 pilots. When five T-38s arrived between May and July 1964, the Deuces were ferried to Perrin AFB for transfer back to the Air Force.

From 1957 to 1960, a pair of Deuces operated from the Ames Research Center at Moffett Field, California. One was used to evaluate the aircraft's auto-attack system, while the other was used mainly to evaluate fire control and auto-maneuvering systems. Other tests were performed with an acceleration command system that could maintain response characteristics throughout all flight modes.

During the early 1970s, a single F-102A was assigned to NASA's Lewis Research Center at Cleveland, Ohio, as a chase aircraft for an NF-106B. The specially modified Delta Dart was assigned to Lewis from 1966 until 1979, with underwing GE 585 engines for supersonic studies of engine inlet designs for the SST. Project work on the SST by the Federal Aviation Administration involved an F-102, which the FAA

This "Tub" was one of four Deuces transferred to NASA's Manned Spacecraft Center at Houston to help astronauts maintain flying proficiency. (NASA)

Astronaut L. Gordon Cooper in flight on 1 January 1963 in one of two TF-102As assigned to the NASA Manned Spacecraft Center. (NASA)

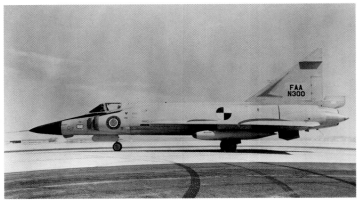

One of the FAA's concessions to the SST program was this F-102A (s/n 57-0835) transferred from the Air Force for the SST program. Test equipment was added below the nose and rear fuselage. Besides the civil registration and FAA emblem, a photo reference disc was added to the center fuselage. (Robert F. Dorr Collection)

In his pre-astronaut days, Virgil "Gus" Grissom poses on 1 April 1960 with an F-102A adorned with four silhouettes that represent Falcon missile firings. (NASA)

acquired on 27 April 1970. Wearing a civil registration, the Deuce was operated from Edwards AFB and the FAA's National Aviation Facilities Experimental Center at Atlantic City, New Jersey. It was transferred back to the Air Force on 15 December and ferried to Davis-Monthan AFB for storage.

In an effort to expedite mandatory F-102 gear loading tests during the summer of 1958, Convair adapted the Navy's mirror landing system used to bring aircraft aboard carriers. By enabling the pilot to maintain the proper glide path through 60 different landing configurations, the tests were completed in only 67 landings versus the 200 expected without the unique system. Test pilot C.E. Myers, Jr., a former Navy pilot, established vertical sink rates from 3 to 10 feet per second. The tests, which wore out 20 sets of tires, also involved single-wheel touchdowns.

In preparation for installation of the General Electric J-85 turbojet engine in Northrop's T-38 "Talon" trainer, the same

Following landing gear tests, this F-102A served as the test bed for General Electric's J-85 engine. (General Electric Aircraft Engines)

Still in FAA markings, serial number 57-0835 rests in desert storage shortly after a 7-month period as a research platform for the SST program. The Deuce retained the red trim of its previous unit, the 182nd FIS of the Texas Air National Guard. (Terry Coxall via Stephen Miller)

Testing at Edwards AFB. The J-85, which was in the 2,000-pound thrust class, was operated at altitudes over 50,000 feet and at Mach 1. Test flights of the J-85-7 aboard the F-102 included afterburner operation to the same altitude and speeds to Mach 1.23.

During fall 1961, a TF-102A (s/n 55-5032) was bailed to North American Aviation's Autonetics Division at Palmdale, California, for the test and development of an automatic approach and landing system. Developed under Air Force contract for the Flight Control Laboratory of the USAF Aeronautical Systems Division (ASD), the APN-114 system controlled the aircraft from the ILS center marker through touchdown. The TF-102A made more than 100 landings at six different airfields during the flight test program. During January 1962, the aircraft and system underwent Air Force evaluation at Wright-Patterson AFB, and from there continued on to the FAA's aviation facility on the east coast for further evaluation. Success with the Autonetics APN-114 led to a commercial version, called "Autoflare," for the Boeing 707.

F-102A was loaned to GE during 1959 for flight testing. Mounted in an instrumented pod beneath the F-102A, the YJ85-5 was put through a 50-hour Preliminary Flight Rating

The F-102A J85 research platform is displayed among GE's high-visibility fleet of test aircraft. Clockwise from the Deuce is a YF-104A, T-38A, XFD-1, RB-66A, F4H-1, and a Sud Caravelle named the "Santa Maria." (Marty Isham Collection)

Mounting the YJ85-5 turbojet, the test F-102A poses with Northrop's T-38 "Talon" trainer, which was to be powered by the new engine. The F-102A's missile bay doors were trimmed to provide clearance for the engine test pod. (General Electric Aircraft Engines)

This boldly marked F-102A, the only NASA-operated Deuce given a NASA registration number, served as a chase aircraft from the Lewis Research Center for an NF-106B mounting a pair of underwing GE J-85 engines. (NASA)

NASA F-102s

Model	Serial	Location	Time Frame
YF-102A	53-1785	Dryden Flight Research Center	9-20-54 to 6-27-58
JF-102A	54-1374	Dryden Flight Research Center	4-3-56 to 5-4-59
F-102A	56-1304	Ames Research Center	4-10-57 to 1960
F-102A	56-1358	Ames Research Center	12-23-57 to 3-21-60
TF-102A	54-1358	Manned Spacecraft Center	9-4-62 to 7-1-64
F-102A	55-3391	Manned Spacecraft Center	9-4-62 to 7-27-64
F-102A	55-3405	Manned Spacecraft Center	9-10-62 to 8-13-64
TF-102A	54-1356	Manned Spacecraft Center	11-1-62 to 7-21-64
F-102A	56-0998	Lewis Research Center	June 1970 to June 1974

Foreign Service

Only two air arms other than the U.S. Air Force operated the F-102: the Air Forces of Turkey and Greece, both of which acquired their Deuces under the provisions of the Military Assistance Program to fulfill NATO commitments. A total of 69 TF/F-102As are known to have passed from the U.S. Air Force inventory to both nations. With the exception of two trainer variants, all were derived from the Air National Guard, with the majority drawn from units in the states of Texas, California, Montana, and Connecticut.

Beginning in June 1968, 22 early model F-102s were passed through the Ling Temco Vought facility at Greenville, South Carolina, for overhaul prior to transfer to Turkey. Most of the balance of F-102s destined for foreign delivery were processed through the Fairchild Industries Aircraft Service Division at Bob Sikes Airport at Crestview, Florida. Seven TF-102As, some of which were requested by both governments as attrition replacements, were ferried directly from their U.S. duty without contractor overhaul.

The actual number of F-102s that were operational with the Turkish Air Force (THK) differs slightly from those earmarked for delivery, since one aircraft (s/n 55-3380) reportedly crashed enroute. Another (s/n 54-1377) is listed contradictory in official records. It is shown as having been destroyed while still assigned to the California Guard, while Turkish documents record its arrival there two months after the crash. The assumption that both records are correct fosters the likelihood that the aircraft was salvageable and shipped to Turkey for use as an instructional airframe.

Both Greek and Turkish F-102s retained their U.S. serial numbers throughout their operational periods abroad. Formerly assigned to the 111th FIS of the Texas Air Guard, this F-102A of the Royal Hellenic Air Force taxis at Tanagra in June 1970. (Stephen Peltz)

The majority of F-102s for foreign service came from the California Air National Guard, including this example, which served the 196th FIS prior to shipment abroad. It is seen here at Murted, Turkey, while assigned to the 144th Squadron in 1970. (Stephen Peltz)

One of six TF-102As turned over to the Turkish Air Force, this Tub arrived in spring 1969. (Norm Taylor Collection)

One of two 1953-vintage Deuces delivered to the Turkish Air Force. (Author)

Following overhaul, the Deuces set out on their delivery flights with stops at the following bases: Goose Bay, Newfoundland; Keflavik, Iceland; RAF Lossiemouth; and Ramstein AB, Germany. The first export Deuces became operational with the Turkish Air Force, beginning in June 1968. The 144th File (Squadron) at Murted received the first batch of F-102s to replace aging F-84s. By late 1969, sufficient numbers of F-102s were on hand to equip a second squadron, the 183rd File at Diyarbakir. Squadron redesignations in 1972 had the 144th File changed to the 142nd, while the 183rd became the 182nd.

Turkey's dispute with Greece over the island of Cyprus during the mid-1970s proved detrimental to the operation of F-102s. Not surprisingly, both sides claimed victory in an air-to-air confrontation that took place in 1974. After Turkish Air Force F-102s downed a pair of Greek F-5s during the engagement, the U.S. imposed an arms embargo on Turkey, severely reducing its ability to maintain its fleet of F-102s. When the 182nd File converted to the F-104 in late 1974, sufficient parts became available to extend the life of Deuces

Deuces of the Greek Air Force were eventually camouflaged in patterns similar to those of their U.S. equivalents. (Jack M. Friell)

of the 142nd File. During June 1979, the last 17 aircraft were withdrawn from service and ultimately scrapped. A total of 35 F-102As and 10 TF-102As served the Turkish Air Force.

Delta Daggers supplied to Greece began arriving in June 1969 to replace F-84Fs. A total of 20 F-102As, along with 4 TF-102As, had been delivered by November 1970. All were

A pair of F-102As of the Royal Hellenic Air Force. (Author)

Having formerly served the Arizona Air Guard, this aircraft was the oldest F-102 in service with the Greek Air Force. (W.H. Strandberg, Jr.)

A small number of Greek F-102s were later painted with a two-tone gray camouflage scheme. Having seen better days, this Deuce ended its service career in a grassy meadow near Tanagra. (John Guillen)

As the Greek Air Force found it increasingly difficult to maintain their fleet of F-102s, this aircraft became an instructional airframe at Ikarus. (Jack M. Friell)

assigned to the 114 Pterighe (Wing) at the HAF Tactical Air Command at Tanagra. Deuces of the Hellenic Air Force flew alongside a variety of surplus U.S. jet aircraft until they were replaced by Mirage fighters in 1978. Like those of the Turkish Air Force, Greek F-102s were either scrapped or found their way into European museums.

F-102s OF THE GREEK AIR FORCE

TYPE	SERIAL NO.	DATE ARRIVED
F-102A	56-0981	Oct 1969
	56-0988	Oct 1969
	56-1001	Nov 1969
	56-1007	Oct 1969
	56-1011	Oct 1969
	56-1016	Oct 1969
	56-1024	Nov 1969
	56-1025	Feb 1970
	56-1031	Oct 1969
	56-1034	Oct 1969
	56-1039	Oct 1969
	56-1040	Dec 1969
	56-1052	Nov 1969
	56-1056	Oct 1969
	56-1059	Dec 1969
	56-1079	Oct 1969
	56-1106	Oct 1969
	56-1125	Oct 1969
	56-1232	Oct 1969
	56-1233	Dec 1969
TF-102A	55-5035	Aug 1970
	56-2326	Jun 1969
	56-2327	Jun 1969
	56-2335	Jun 1969

F-102s OF THE TURKISH AIR FORCE

TYPE	SERIAL NO.	DATE ARRIVED
F-102A	53-1814	Oct 1968
	53-1815	Jul 1968
	54-1377	Jun 1968
	54-1379	Sep 1968
	54-1380	Jul 1968
	54-1382	May 1969
	54-1383	Oct 1968
	54-1384	Sep 1969
	54-1403	Nov 1968
	55-3383	Jun 1969
	55-3384	Sep 1969
	55-3385	Jul 1969
	55-3386	Jan 1969
	55-3389	Sep 1969
	55-3390	Sep 1969
	55-3392	Oct 1968
	55-3395	Jul 1968
	55-3396	May 1969
	55-3400	Jan 1969
	55-3403	May 1969
	55-3404	Sep 1968
	55-3405	Oct 1968
	55-3408	Jun 1969
	55-3409	Sep 1969
	55-3410	Nov 1968
	55-3412	Sep 1968
	55-3416	Sep 1969
	55-3420	Jun 1969
	55-3421	Sep 1969
	55-3426	Jul 1968
	55-3429	Jun 1969
	55-3446	Jun 1969
	55-3452	Jan 1969
	55-3455	Sep 1969
	55-3461	May 1969
TF-120A	54-1352	Jun 1968
	54-1360	Jun 1968
	55-4032	Aug 1970
	55-4033	Apr 1969
	56-2325	Dec 1970
	56-2334	Nov 1970

William Tell

Centuries ago, a man named William Tell rescued his homeland from a ruthless tyrant through superior skill, strength, and courage. In a modern version of William Tell, the U.S. Air Force tested those same qualities of its air defense system with a worldwide weapons meet. The Air Force adopted the name since William Tell's legendary ability to pierce an apple with an arrow seemed the appropriate theme of the weapons meet.

The program called for deployment of an interceptor unit complete with aircraft, weapons, and maintenance personnel to a site where they were pitted against targets in a realistic combat scenario. As a continuous evaluation of tactics and weapons systems, the meet recognized the best aircrew-controller team in the air defense system, rated crews that maintained and loaded aircraft weapons, and demonstrated the capabilities of USAF interceptor weapon systems.

The competition originated in 1952 as the U.S. Air Force Gunnery and Weapons Meet, in which aircraft fired at towed banner targets. A Fighter Interceptor/Air-to-Air Rocketry Phase was introduced during the 1954 meet, in which only four teams participated: three from the Air Defense Command and one from the Air Training Command. Competing in the event were classic jets of the period: the F-94C Starfire and the F-86D Sabre. Adding to the competitive flavor of the 1955 meet were four teams from overseas USAF bases, which doubled the number of teams that participated the previous year. Nine teams, representing every major command concerned, competed in the 1956 Interceptor Phase of the competition. The event also marked the end of banner targets in rocketry competition. Thereafter, radio-controlled drone targets and an electronic scoring system would change the face of William Tell. The 1957 meet was not held, since all training was in the process of being moved to Vincent AFB.

Yuma Proving Ground, along with Vincent Air Force Base and the 4750th Air Defense Wing (Weapons) had served as hosts for the first four air defense meets, while the 4756th ADW (Weapons) would host succeeding competitions at Tyndall Air Force Base. Located on Florida's coast, Tyndall's proximity to the Gulf of Mexico proved ideal for overwater range operations. Meets lasted ten days and were held in September or October.

The 1958 competition, which was the first William Tell Weapons Meet, deviated dramatically from all previous events. Besides featuring the first competitive use of the F-102, the meet was divided into separate categories for the first time due to the vast differences in attending aircraft and weapons. The event also saw the first competitive use of the Falcon and Genie missiles, plus the introduction of closed-circuit television to monitor both air and ground activities. The 1958 meet marked the final appearance of the F-86 fighter in weapons competition.

The selection of squadrons and crews to participate in William Tell was usually based upon overall unit performance and the results of competitions held by parent commands. The rules and scoring of such events paralleled those of the actual meet. Besides pilots, the normal compliment of a team participating in William Tell comprised two controllers, forty airmen and officers, and three technical representatives. During the 1958 meet, a total of 16 sorties was flown by each team, with four pilots each flying four missions. Of the 16 sorties, 12 were live fire with missiles fired at Firebee target drones. Of the four sorties per pilot, two were flown above 40,000 feet, one below 1,000 feet, and one to test the pilot's skill in electronic countermeasures.

Testing the effectiveness of America's long-range air defense system during the Fighter Interceptor Phase of the 1959

A pair of Wisconsin Air National Guard F-102As are surrounded by ground support equipment as they are loaded with missiles at Tyndall AFB in 1972. (WI ANG)

Believed to be a non-operational airframe, this Deuce served in the 1972 William Tell competition. The name "Dyna" was painted on the nose landing gear door. (Lionel Paul Collection)

Poised on Tyndall's flight line for the 1958 William Tell are F-102s, F-86s, and F-89s from 12 teams. Visible in the right background are B-26 drone launch aircraft, Firebee drones, and an H-21 drone recovery helicopter. (USAF)

Atop a Q-2A Firebee target drone at the 1958 weapons meet, a jet age model of the legendary William Tell poses with her bow, the arrow a miniature version of the Falcon missile. (USAF)

Project William Tell were six F-102A, three F-89J, two F-104A Starfighter, and one F-100A Super Sabre squadrons. Both the infrared, heat-seeking GAR-2, and GAR-8 missiles made their debut at the meet. Other firsts included a multiple drone mission and a night sortie for each team.

A typical mission had a B-26, with a Firebee mounted under each wing, climb to altitude over the Gulf of Mexico missile range. Soon after a drone was released to begin its track, a weapons controller picked up the target on his radar scope and scrambled the interceptors. Quickly airborne, the jets were vectored into position by ground controllers until the pilots' radar latched on to the target. The interceptors then closed for the kill, at which point electronic scoring systems activated to score the firing results. Points were awarded not

only for hits, but also near misses, since those recorded against the diminutive drone would have destroyed an enemy bomber. The performance of weapons loading teams and ground controllers was also scored.

William Tell's apple took the form of the elusive Ryan Q-2A Firebee target drone. The jet-powered drones provided realistic targets by flying at speeds over 600 mph, above 50,000 feet, and for durations exceeding one hour. After release from the mother ship, the craft was flown by radio control from a ground station. Besides the ability to transmit hits and misses to scorekeepers on the ground, Firebees could be equipped with cameras to record inbound missiles and rockets, flares to draw heat-seeking missiles away from its engine, and radar augmentation to simulate large aircraft. If

Pilots and ground crew of the 482nd FIS pose during the 1958 William Tell, while the welcoming committee graces one of their aircraft flown in the meet. (USAF)

A Deuce of the 4750th Test Squadron fires a GAR-1 Falcon missile during the 1958 Worldwide Weapons Meet at Tyndall AFB. (USAF)

An F-102A, loaned to the 325th FIS by the 48th FIS for the 1959 William Tell, rotates on takeoff from Tyndall during the competition. (Marty Isham Collection)

Team members of the 482nd FIS pose with "Bear Grease" (s/n 57-0840), one of the squadron's entries in the 1958 Worldwide Weapons Meet. The 482nd was selected to compete in the succeeding competition, as well. From left to right are A/2C Barnes, A/2C Wagner, A/2C Markley, S/SGT Griffin, and Capt. Stacey. (USAF)

the drone survived, it was flown to a pickup point, shut down, lowered by parachute, and retrieved from the ocean by boat or helicopter.

After 1959, meets were held every two years. Although F-102s of the Spain-based 497th FIS had been selected to represent USAFE in the 1961 competition, overseas participation was canceled due to the Berlin Crisis. Noteworthy at the 1963 meet was the first international team to attend a William Tell, formed by the 4th FIS, representing PACAF, and personnel from the Japanese Air Self-Defense Force. Combat commitments in Southeast Asia—in a sense, a long-term,

enormous version of William Tell, conducted in earnest—caused the cancellation of the 1967 and 1969 meets.

Air National Guard F-102 teams participated in six William Tell competitions, beginning in 1961. The 1970 meet was the first to feature all National Guard F-102 units. The 1974 competition was the last in which F-102s participated as interceptors. From then on, until William Tell 1984, they would appear at Tyndall as full scale target drones.

Clearly present in William Tell's competitive environment was the bravado and bluster that personified the jet fighter community. Rivalry was at its peak, as teams from across the

Winners of the 1972 William Tell, F-102 pilots of Wisconsin's 176th FIS pose with the markings specially designed for their aircraft during the event. All wear patches that commemorate both their home state and the Air Force's 25th anniversary. From left to right are Lt. Col. Brickson, Maj. Laquey, Capt. Foster, Capt. Ebben, and Maj. Manthey. (USAF)

Lt. Col. Brickson, Team Chief of the 176th FIS for the 1972 William Tell, poses with the trophy awarded the unit for First Place in the F-102A category. (WI ANG)

globe battled for first place in their category and the coveted first place overall trophy. Realism was the keynote.

Besides keeping air defense units at knife-edge performance, William Tell served other purposes. Equally important was its role in showcasing the latest in aircraft, weapons, and electronic hardware. Drawn to the event were thousands of visitors, many of whom were scientists, high-ranking military officials, foreign dignitaries, politicians, and representatives from the aerospace industry.

William Tell, with the F-102 as a key player, had sent a subtle but clear message to a world preoccupied with Cold War that America's air defense capabilities were second to none. That premise thrives in today's William Tell—called the Worldwide Air-to-Air Weapons Meet—its mission as vital to training and evaluation as when Korean war vintage and century series jets gathered at Tyndall.

F-102A PARTICIPATION IN WILLIAM TELL

EVENT YEAR	UNIT	HOME BASE	OTHER PARTICIPATING AIRCRAFT
1958	*326 FIS	Richards-Gebaur AFB, MO	F-86D/L, F-89J
	482 FIS	Seymour Johnson AFB, NC	
	317 FIS	Elmendorf AFB, AK	
	318 FIS	McChord AFB, WA	
1959	326 FIS	Richards-Gebaur AFB, MO	F-89J, F-100A, F-104A
	482 FIS	Seymour Johnson AFB, NC	
	317 FIS	Elmendorf AFB, AK	
	*460 FIS	Portland, OR	
	525 FIS	Bitburg AB, Germany	
	16 FIS	Naha AB, Okinawa	
1961	*59 FIS	Goose AB, Labrador	F-101, F-106
	182 FIS	Kelly AFB, TX	
	317 FIS	Elmendorf AFB, AK	
	331 FIS	Webb AFB, TX	
	3558 FIS	Perrin AFB, TX	
1963	59 FIS	Goose AB, Labrador	F-101, F-106
	*146 FIS	Pittsburgh, PA	
	317 FIS	Elmendorf AFB, AK	
	4 FIS	Misawa AB, Japan	
	525 FIS	Bitburg AB, Germany	
	460 FIS	Portland, OR	
1965	*32 FIS	Camp N. Amsterdam, Neth.	F-101, F-106, F-104
	59 FIS	Goose AB, Labrador	
	326 FIS	Richards-Gebaur AFB, MO	
	64 FIS	Paine Field, WA	
	157 FIS	McEntire ANGB, SC	
1970	*148 FG	Duluth, MN	F-101, F-106
	142 FG	Portland, OR	
	124 FG	Boise, ID	
1972	*115 FG	Truax Field, WI	F-101, F-106
	57 FIS	Keflavik NAS, Iceland	
	158 FG	Burlington, VT	
1974	*124 FIG	Boise, ID	F-101, F-106
	112 FIG	Pittsburgh, PA	

* Top Team in Category II - F-102A

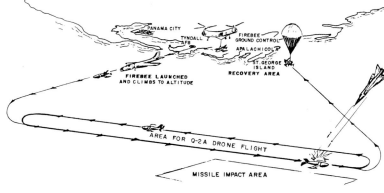

William Tell course at Tyndall AFB, FL.

Final Flight - Deuce Drones

Since the F-102 routinely scrambled from bases across the globe to confront the unknown, a takeoff from Holloman AFB, New Mexico, to fly a profile over the neighboring White Sands Missile Range would seem routine. But operations from Holloman were different. It was over White Sands that F-102s were to meet their prearranged fate—an attack by an Army surface-to-air missile or an Air Force air-to-air missile. Similar scenes were played out over the Gulf of Mexico where F-102s, having launched from nearby Tyndall AFB, Florida, were attacked by fighters firing air-to-air missiles. Scores of America's first supersonic jet interceptor would become target drones in scenarios where the hunter became the hunted.

Prior to the use of F-102s as drone aircraft, missiles were tested primarily against the "Firebee" family of sub-scale drones. However, they lacked the maneuverability of full-scale types and had to be augmented with systems that reproduced the infrared or radar signals of real aircraft. There were a number of full-scale target programs from the mid 1940s to the U.S. Air Force's QF-104 program during the early 1960s. Before escalating costs brought the QF-104 program to an end, it proved that full-scale drones were final proof of weapons system design. When first tested against the QF-104, missiles such as the AIM-9J and AIM-7E, which were effective against sub-scale targets, passed harmlessly behind the Starfighter.

Those embarrassing findings paralleled data in the USAF Red Baron Report of aerial engagements in the Vietnam war. The hiatus following the QF-104 program had a detrimental effect on the air war. Although American pilot skill prevailed, air-to-air missiles fired at enemy aircraft did not always perform as expected. Although U.S. Air Force weapons had been thoroughly tested against sub-scale drones, heat-seeking missiles often exploded in the afterburner plume of Migs, and the number of actual radar missile kills fell far below acceptable levels. Such glaring drawbacks gave rise to a study for a new line of full-scale targets.

At its conclusion, the requirements were presented to the Air Force Chief of Staff in February 1972, prompting the decision to convert aging F-102s into target drones. The Deuce was the obvious choice. Besides the fact that nearly 400 F-102As were in storage, the Deuce's characteristics best duplicated those of threat aircraft by providing realistic performance, size and radar, infrared, and visual signatures.

Edging out six major contractors, the Sperry Rand Corporation was awarded the Air Force contract on 1 April 1973 for the modification of F-102As into two different drone configurations. Initial QF-102As retained pilot controls for contractor-operated flights, while those designated PQM-102As were "de-man-rated," and the first full-scale targets with afterburners. Sperry had 24 years of experience in converting manned aircraft to drones, beginning with a program in 1945 to drone F-80s for the Air Force. Its success led to drone conversions of B-17, T-33, B-47, and F-104 aircraft. Prior to the Air Force contract, the firm had conducted internal studies of the unmanned PQM-102 concept.

Sperry began work on the drones at their facility in Crestview, Florida. An initial contract called for three QF-102As, one of which made the first flight on 10 January 1974. A follow-on contract for two QF-102As was followed by 65 PQM-102As and another 145 as PQM-102Bs. The first flight of the PQM-102A, which was the first unmanned flight, took place at Holloman AFB on 13 August 1974. Beginning in 1975, Sperry was producing four aircraft per month. At the start of PQM-102B production in 1978, the line was transferred to Sperry's Hangar 52 at Litchfield Park, near Phoenix, Arizona. When production ceased in 1981, a total of 215 F-102As had been converted.

The U.S. Air Force and Sperry Flight Systems, along with contract partner RCA, worked jointly throughout the F-102 drone program, which was given the name PAVE DEUCE. The interaction among the unique blend of civilian contractors, many of whom were former Air Force members, and

Having seen better days, this Deuce was one of the first QF-102A conversions and the first of two later modified to QF-102B standards. (Terry Love)

Wearing fresh paint and carrying an electronics pod under its wing, the first QF-102B taxis to the "droneway" at Tyndall AFB in 1982. (Author)

Still wearing its Florida Air Guard scheme, the first aircraft for the PQM-102B modification program (s/n 56-1077) arrives at Litchfield in May 1978. (Honeywell)

Named "La Tina," this TF-102A was transferred to Tyndall AFB during June 1975 for crew training in the drone program. It was placed in storage during August 1982 and turned over to a private museum in September 1983. (Terry Love)

military personnel, played a large part in the success of the PAVE DEUCE program. Development and management of the PQM-102 series was overseen by the Armament Development Test Center at Eglin AFB, Florida. At Tyndall AFB, the actual operation of the drones was handled by the 4756th Test Squadron. The 6585th Test Group of the Air Force Special Weapons Center at Holloman was responsible for system trials on the White Sands Missile Range. Beginning in July 1981, the 82nd Tactical Aerial Targets Squadron was activated at Tyndall to operate the Air Force drone fleet. The unit's Detachment One was established at Holloman to support the Army's missile test program.

The drone conversion process began at the Military Aircraft Storage and Disposition Center (MASDC) at Davis-Monthan AFB, Arizona, where the aircraft were stored. They were made flyable and test flown by MASDC pilots before delivery to Sperry. There they were defueled, drone markings were applied, and the aircraft were disassembled and inspected. The Deuce was stripped of radar, fire control, and

navigation gear to make room for the new avionics. Hydraulics were overhauled, and the engines were removed and unused wiring stripped out. A new wiring package was then fabricated to precisely match the location of each new component. Next came the installation of electronic flight control systems, command and telemetry systems, automatic destruct, scoring, and other special remotely and automatically controlled systems that produced full-scale aerial targets.

Included in the conversion were modifications made to the airframe: fuel control, brakes, landing gear, afterburner, throttle, canopy, antenna array, and electrical wiring. The specific equipment package installed and how it was arranged depended on whether the aircraft was to be flown remotely or by a pilot.

When flown by a pilot, the aircraft was designated a QF-102A, which had its electronic pallet assembly installed aft of the cockpit. In the remote mode, the assembly occupied the cockpit and the designation PQM-102A/B was used. Designations were used interchangeably since the aircraft required

The first four Deuces of the PQM-102B program undergo the conversion at Sperry's Hangar 52 at Litchfield Park in 1978. (Honeywell)

Seen here with the first PQM-102B, the pre-mission test stand was used to check flight phases prior to actual flight operations. (Honeywell)

No longer needed, the fire control system was removed from the forward electronics compartments to make room for drone-related control systems. (Honeywell)

Originally modified as a QF-102A, this Deuce was later designated a QF-102B for use as a man-rated demonstrator. It is seen here on display at Tyndall AFB for the 1982 William Tell competition. The scoreboard at left lists all participating units. (Author)

a pilot for ferry flights to range sites, missile evaluations, training, system checks, and transfers between operational sites. Reconverting a PQM-102 to accommodate a pilot was a relatively simple process since the electronics pallet was positioned on the ejection seat rails.

The PQM-102B was a low cost version of its predecessor with a savings of $100,000 per aircraft. The QF-102As were assigned aircraft numbers beginning with 601, except for two numbered 501 and 502. Eventually configured as PQM-102Bs, the pair was redesignated QF-102Bs since they were flown with a pilot on a regular basis. All PQM-102Bs were numbered from 701 onward. Unmanned drones were labeled "Nullos," a Greek term meaning "No man," which became popular in the drone community. Once operational, the Deuce drones took on the call sign "Spad," which was shortened from "Sperry Aircraft Delivery."

The Deuce drone had a basic weight of 20,000 pounds and an operational mission weight of 31,200 pounds. A payload capability of 4,200 pounds was available using the aft electronics bay, missile bay, cockpit, and two wing pylons. Standard payloads included chaff, flares, and a wide variety of electronic countermeasures equipment. The aircraft could attain a speed of Mach 1.2 and operate at altitudes between 200 and 56,000 feet. It could withstand 8 Gs and fly at a range exceeding 200 miles, depending on the effective control range of guidance radar. Remote control of the Deuce drone—along with all full-scale Army and Air Force targets used during the 1970s and '80s—was accomplished with the Vega-developed Drone Tracking Control System and the Army-sponsored/IBM-developed Drone Formation Control System. The latter used mainframe computers at the ground station to control more than one drone if a multiple threat

Deuce drones were equipped with a smoke generating system located in the center fuselage. A chemical was piped to a streamlined nozzle at the tail, where it was sprayed into the exhaust to produce smoke. (Author)

The drone's nemesis—an AIM-4D (left) and AIM-4G Falcon missile were part of the hardware display during the 1980 William Tell competition at Tyndall AFB. (Author)

Deuce drones await their turn on the Gulf range at Florida's Tyndall AFB. (USAF)

An inbound missile is captured on film milliseconds before it impacted a Deuce drone. (Honeywell)

scenario was desired. Scoring systems included the DIGIDOPS system by Cartwright, and later, the Motorola Vector Miss Distance Indicator System. Two types of scoring cameras provided fore and aft coverage of missile approach angles, speeds, and miss distances. A smoke-generating system provided visual means of identifying the target to ground cameras and attacking aircraft. A smoke trail also made the drone visible to chase aircraft closing to check missile damage that may have ruled out a safe landing. A vital component incorporated into the PQM-102 was a dual destruct system that utilized a MK-48 explosive warhead. When detonated, the 25-pound charge, located in the center missile bay, could literally cut the aircraft in two.

When modifications were completed, the aircraft underwent a thorough check of remotely controlled electronics and electrical circuitry by a pre-mission test stand. Nearly all flight phases that ground controllers used to guide the drone could be simulated during static tests. Preparations were exhaustive. For every mission, an average of three manned flights were made, and more than 100 system tests took between 12 and 46 hours to complete. Prior to delivery, a test flight for airworthiness was made by a pilot who very likely flew the type during Air Force service. Previous flight time in the F-102 was a specific job requirement to be a drone test pilot since the type of flying was very demanding. Low level missions with 5 G turns left little room for lack of coordination or experience. Veteran Deuce pilots were quick to admit that their reflex actions were tempered when flying the aircraft from the totally different perspective of a ground control panel. Control systems of U.S. Navy QF-86 target drones and Japa-

The missile hit seems to have assured the PQM-102B's destruction. (Honeywell)

Some drones that received superficial damage were flown remotely back to bases for repeated use. Serial number 56-1418 was recovered after a missile hit damaged its speed brakes in 1974. It flew seven drone sorties before it was destroyed at Holloman AFB. (USAF)

One of the few QF-102As, this aircraft was assigned to the Air Defense Weapons Center and based at Tyndall AFB. During the 1980 William Tell meet it wore nose art in reference to the then-popular "Dallas" television series. (Norm Taylor)

Wearing a variety of schemes and markings indicative of their former air defense assignments, F-102As undergo the conversion to PQM-102Bs during late 1979. (Honeywell)

nese QF-104s differed in that they used an onboard TV camera, which enabled a ground control pilot to fly the aircraft as though he were in the cockpit.

At the range, the aircraft was prepared for what could be its final flight. It was towed to the runway, the destruct system armed, and the engine started. Beside the runway were two RCA controllers who controlled the drone from atop a mobile control station. One handled the power and pitch control, while the other controlled the heading with ailerons and rudder. After liftoff, control was passed to two pilots at a fixed station who flew the target into a pattern and maneuvered it during missile firings. The drone was controlled throughout its flight by a long range command radar link with a separate link used for backup control. On command from the ground site, the drone could automatically execute a series of severe evasive maneuvers and throw electronic or infrared countermeasures at the attacker. A normal mission lasted from 40 to 55 minutes, depending on afterburner use and altitude. If the

drone survived the flight, control was passed back to the pilots near the runway.

Seldom were missiles armed with warheads since the idea was to prolong the drone's use. Therefore, direct hits were necessary to destroy the drone. Since missiles would destroy an enemy aircraft if they exploded within a kill circle, training crews were credited with a kill within a scoring distance. Some missiles used proximity fuses, while others carried telemetry gear to send data back for scoring analysis. If the shooters did not physically hit the target or inflict major damage, and the drone was controllable, operators returned it to the base. The PQM-102's airspeed was reduced so a chase plane, usually a T-33 or T-38, could move in to inspect the damage before it was handed off to the mobile control station.

At no time during the flight was there danger of a drone becoming errant and threatening populated areas. Immediately after takeoff from Tyndall, a drone was over the Gulf of

This F-102A went from the San Antonio AMA to the drone program in April 1973, where it became a QF-102A with the number 602. Unusual is the non-standard camouflage and large national insignia. (Terry Love)

Initially assigned to the 57th FIS, this Deuce also served the Idaho and California Air Guards before transfer to the drone program during December 1975. It survived 48 missile firings during 16 sorties before its demise on 17 November 1981. (John Guillen)

This Deuce underwent the transformation to PQM-102B during late 1978 and was destroyed at Holloman on its second drone sortie on 3 August 1981. A corporate emblem bearing the acronym "SNORT" was placed over the center of the national insignia. (Terry Love)

Few F-102s wore the ever-popular shark's mouth. This PQM-102B crashed on takeoff at Holloman AFB on 16 July 1980 after having flown only one sortie. (Terry Love)

Mexico, where its mission pattern was flown 50 miles offshore. Had the target become sufficiently damaged or uncontrollable, the onboard explosives could be detonated several ways; or it could be destroyed by simply flying it into the ocean. The fail-safe destruct system was especially important at White Sands Missile Range, where airspace restrictions were more severe. An alternate 35,000-foot strip was also available to land a crippled drone.

A kill ratio of one out of four ensured that most Deuce drones ended their years of yeoman service in a fiery explosion. Many were downed on their first flight, while others returned to the skies to challenge their attackers. The record survivor was PQM-102B number 734 (s/n 57-0806), which endured 20 missions before it was destroyed over the Gulf of Mexico. Taking first place in the defiance category was number 644 (s/n 56-1426), which survived 48 missile hits during 16 missions before its demise at Holloman.

Early in the flight test phase of the PQM-102A, the value of the program became evident when missiles, that were developed against hot spot simulators, failed to hit the full-scale target. When Sperry took the pilot out of the cockpit and converted the F-102 to drone status, it gave pilots and ground gunners the opportunity to train against a full-size, highly maneuverable, afterburning target—very similar to the Soviet SU-19, MiG-23, and MiG-25. The Deuce didn't fly in the Mach 3 range as did the MiG-25, but its radar and infrared image was basically the same. The F-102 was a star performer in the drone role because of its capabilities. It was one of the best supersonic airplanes, especially at low altitudes, and though it was beyond its years, was fairly easy to maintain.

Not limited to target use, it served the Air Force in tactics and weapons training, ECM tests, research and development of new missiles, pilot training, and weapons delivery testing.

Its upper surfaces painted black with red trim, this PQM-102B carries a blue electronics pod on the wing pylon at Tyndall in 1982. (Author)

Still in its distinctive Pennsylvania Air Guard scheme, this Deuce drone was destroyed at Holloman's range on 29 August 1978. (Author)

During its air defense career, this Deuce wore serial number 56-1400. It was the third drone built, having entered the program during April 1973. To showcase the success of the drone program, it was given this distinctive scheme, complete with Sperry and RCA emblems. (Terry Love)

Photographed from a chase aircraft, a pilotless PQM-102A flies over the Holloman range in 1979. Markings on the weather-beaten drone included a large, yellow stenciled "8" on the radome, a large pumpkin face on the center fuselage, and the remnants of stars on the rudder, revealing its past with the California Air Guard. (USAF)

Deuce drones were regular participants at the William Tell Weapons Meet. Air-to-air missiles tested against the PQM-102 included the AIM-7E/F Sparrow and AIM-4E Falcon beam riders. Having procured 14 Deuce drones, the U.S. Army made extensive use of the type for its missile test program at White Sands. Army surface-to-air missiles tested against the PQM-102 were the Stinger, Roland, Chaparral, and Patriot. The PQM-102s at Tyndall were also used to evaluate the McDonnell Douglas F-15 "Eagle" air-to-air weapons systems.

Of particular interest was the mission flown by serial number 57-0909, the last F-102 manufactured by Convair. Removed from desert storage, number 909 was put through the drone conversion process and sent aloft as a target on 24 February 1982. It returned with hits from missiles fired by F-4s recorded, but had no major damage. On 3 March, number 909 went up against the Michigan Air National Guard's 171st FIS flying F-4s. Among the "shooters" was Lt. Col. Arendt and his Weapons System Officer, Capt. Cumiskey, in Phantom II serial number 63-7412. They lined up on the drone and launched an AIM-7E missile, which blew away 909's afterburner shroud. The ground controller was able to bring the crippled drone back on station, allowing another Phantom to deal the death blow with a second AIM-7E. Trailing flames, the drone continued to fly, making it necessary for ground control to activate the destruct system. Few people then realized the irony of the fact that Col Arendt's Phantom, the

Trailing a fiery wake, a Deuce drone falls from the sky after a missile hit. (USAF)

Easily identified as having served with the 57th FIS, this F-102A was used to test the drone's MK-48 explosive warhead at Holloman AFB in July 1974. This view, shortly after detonation, is testimony to the warhead's ability to shear the aircraft completely in half in the event a drone became uncontrollable. (Honeywell)

Under the watchful eye of technicians in the mobile control station, one of the first PQM-102As makes a controlled departure from Holloman in 1973. The dome-shaped antenna atop the van was command and telemetry radar, while the shorter one was omni radar. (Honeywell)

Its cockpit eerily pilotless, the second built PQM-102A soars at altitude. It completed nine sorties before it was destroyed over Tyndall's Gulf range in 1975. (Honeywell)

USAF's oldest operational F-4, which was assigned to the first ANG air defense unit to fly Phantoms IIs, had shot down the last produced F-102.

The last F-102A for the PAVE DEUCE program was flown from the storage facility on 29 July 1981. As F-102A stocks were depleted, it was decided to convert the F-100 Super Sabre, nearly 400 of which were in storage. The development program to transform the "Hun" to a QF-100 full-scale

target drone began in September 1979. Deuce drone operations came to an end at Tyndall on 14 July 1983, when Spad 793 was guided out over the Gulf of Mexico for a fateful rendezvous with four F-4 Phantoms. However, the PQM-102B sustained only slight damage to a wing and was returned to Tyndall. On 9 August 1983, Spad 793, along with another drone remaining at Tyndall, were flown to Holloman, where the final PAVE DEUCE mission was flown on 30 June 1986. The PQM-102A/B flew a total of 760 missions.

A PQM-102B flies off the wing of its predecessor, the QF-100. Both have pilots in the cockpit, a common occurrence during test profiles in the event ground control went awry. The QF-100 carries an electronics pod on the left wing pylon. (Honeywell)

The PQM-102B number 782 shares the ramp at Sperry Flight Systems in Arizona with other drone aircraft in 1980. From left to right are the Navy QT-38, Navy QF-86, and Japanese QF-100. (Honeywell)

Not only are present day drone aircraft less costly than sub-scale types, they are supersonic, carry more payload, and are capable of 9 Gs. They have sophisticated control systems that do not require a ground control station. Although their days as front-line aircraft are over, they serve a vital function in the unmanned role by ensuring that the Free World's missiles are accurate, reliable, and as capable as possible. They have provided air crew and gunners with thousands of realistic duplications of fast, highly maneuverable, afterburning-equipped opponents they are likely to confront in future conflicts.

Sequence views of F-102 drones during their final flight. (USAF)

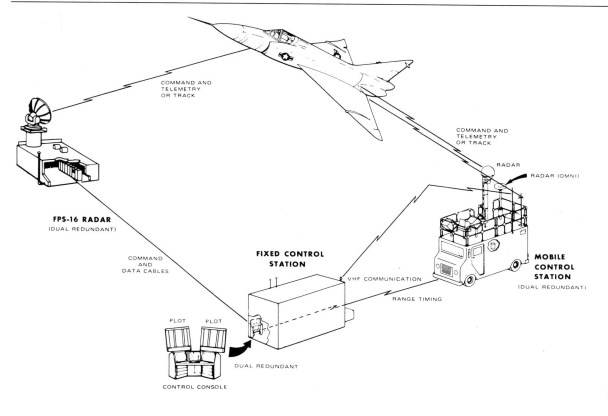

COMMAND AND
TELEMETRY
OR TRACK

COMMAND AND
TELEMETRY
OR TRACK

RADAR

RADAR (OMNI)

FPS-16 RADAR
(DUAL REDUNDANT)

COMMAND
AND
DATA CABLES

**FIXED CONTROL
STATION**

VHF COMMUNICATION

RANGE TIMING

**MOBILE
CONTROL
STATION**
(DUAL REDUNDANT)

PLOT   PLOT

DUAL REDUNDANT

CONTROL CONSOLE

PQM-102 Target System

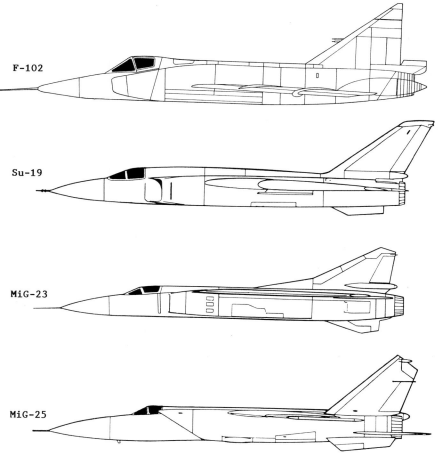

F-102

Su-19

MiG-23

MiG-25

These profiles illustrate the PQM-102's effective representation of threat aircraft.

F-102 Survivors

Although most F-102s were eventually scrapped or met their fate on target ranges, a substantial number survived to become "gate guards," or museum displays. The majority of Deuces made their final flight to the Military Aircraft Storage and Disposition Center (MASDC), now called the Aerospace Maintenance and Regeneration Center (AMARC), which adjoins Davis-Monthan AFB, near Tucson, Arizona. As of 1998, only four Deuces remained at the sprawling site: three F-102As (s/n 56-1455, 56-1515, and 57-0821) and one TF-102A (s/n 56-2350). All were flown to the storage facility during the early 1970s, with the exception of 56-1515, which arrived there during 1998.

More than 70 F-102s went on to serve as monuments in tribute to the interceptor's role as Free World guardian. Some Deuces were earmarked for preservation because they distinguished themselves in some way. Such is the case with the F-102A exhibited at the U.S. Air Force Museum at Wright-Patterson AFB, Ohio, where number 56-1416 was flown in 1971. Having served with the 57th FIS in Iceland, number 416 was one of the first USAF aircraft to intercept and escort a Soviet "Bear" bomber over the Arctic. On a similar note, the pedestal-mounted Deuce at NAS Keflavik, Iceland, serves as a memorial to the personnel of Air Forces Iceland, who formed a vital link in NATO defenses. During December 1969, the familiar red tails and delta-shaped wings, which had been a part of the Alaskan scenery for more than 12 years, lifted from the runway at Elmendorf AFB for the last time. One F-102A of the Alaska Air Command's 317th FIS stayed behind to be towed through the streets of Anchorage to its new home at the Alaska Transportation Museum.

Some facilities are in possession of more than one Delta Dagger, often as a result of the Air Force Museum's loan program. Thanks to the museum's preservation efforts, a number of Europe-based F-102s were spared destruction. It was also not uncommon for some F-102s to have been displayed at more than one site. An F-102 that served as a gate guard at Syracuse, New York, for 17 years was acquired by the Minnesota Air National Guard in 1988. After more than five years of complex restoration, the badly deteriorated Deuce was transformed into an impressive display at the Minnesota Air Guard Museum.

Not all Deuces started their static career with their original serial number. Some were given a different number to commemorate aircraft and crews lost in crashes, or an aircraft once flown by the tenant unit where the aircraft is displayed. For example: though serial number 56-1032 was converted to a drone and destroyed, the Dutch Zeist Museum placed an F-102 on display in 1998 bearing the serial number 56-1032. The museum chose to represent that particular aircraft in view of its significance in the Holland-based 32nd FIS. It served not only as the commanding officer's mount, but bore the squadron's numerical designator within its serial number.

Since some F-102s converted to drones were not destroyed, they were put to use in other programs that sealed their fate. Four or five arrived at Aberdeen Proving Ground, Maryland, in 1989 for use as ground targets in Army anti-aircraft munitions tests. They were flown to the site since Aberdeen's 12,000-foot runway could accommodate them.

Of special interest are a number of F-102s that were acquired by the New Mexico Institute of Mining and Technology at Socorro, New Mexico. They were used for ordnance tests by the Energetic Materials Research Test Center of the college, which was under contract with the U.S. Navy's Lethality Weapons and Effectiveness Group of the Analysis Center Group at Dahlgren, Virginia. The selection of the F-102 for the tests was based upon its design similarities to potential adversary aircraft. The airframes, minus their engines, were

Having ended its career as an instructional airframe with the Connecticut Air Guard, this beautifully preserved Deuce is displayed at Bradley Airport, Connecticut, with markings from its days as a commander's aircraft with the 525th FIS in Europe. (Lennart Lundh)

Sealed for protection from the elements, F-102s await their fate in desert storage near Tucson, Arizona, in 1979. (Terry Love)

Having been resurrected from desert storage, this F-102A, formerly of the Wisconsin Air Guard, was moved to the MASDC flight line in 1979. Protective screens were attached to the air intakes during its preparation for the drone program. (Terry Love)

Though having been in storage for more than two years when this photo was taken, the markings of this Tub clearly identify its last assignment with the 57th FIS in Iceland. (Larry Davis)

Wisconsin's Volk Field boasts an F-102 formation display comprising both primary types. Volk was a major training site during the Deuce's heyday in the Air National Guard. (Author)

Wearing the bold markings of the 68th FIS, this former Florida Air Guard Deuce (s/n 56-0986) was placed on display at Florida's Clearwater Airport. (Author)

Like a number of Deuces, this example was retained for display purposes where it served as an Air Guard interceptor. After service with the Arizona Air National Guard, this F-102A assumed gate guard duty at Tucson International Airport. (Gregory Spahr)

One of the first F-102As is currently displayed at Joe Foss Field, Sioux Falls, South Dakota, marked as a Deuce once flown by the South Dakota Air Guard. Its original serial number appears at the base of the vertical tail. After serving with the California ANG, the actual 56-1114 resides at the March Field Museum in California. (Jeff Kolln)

In markings familiar to the California Air Guard, its last duty assignment, serial number 56-1114 graces the grounds at March AFB. (Gregory Spahr)

The Pima Air and Space Museum's vast historic aircraft collection includes this TF-102A, which last served the California Air National Guard. It is one of the few preserved in Vietnam war era camouflage. The Pima collection also includes an F-102A. (Dale Mutza)

This TF-102A poses near the main gate at Kelly AFB, Texas, where Deuces were regularly assigned and overhauled. (Author)

One of only two YF-102As known to remain in existence, this well-maintained Deuce rests among a wide variety of aircraft on display at Fort Walton, Florida. The emblem of the then-tenant unit, the 1st Special Operations Wing at nearby Hurlburt Field, was applied to the aircraft's nose. The F-102A was previously displayed at McEntire ANGB, South Carolina. (Lionel Paul)

This TF-102A ended its career as a test/chase aircraft for the Air Force Flight Test Center as a permanent display at California's Fox Field. (Author)

Set against the Alaskan scenery, this impressive F-102 display stands at Elmendorf AFB, Alaska, in tribute to the Alaskan Air Command. (Author)

After serving with the ADC, AAC, USAFE, and the ANG of three states, this Deuce ended its long service life at South Carolina's McEntire ANGB. (Terry Love)

Once used to test the SAGE system, this early F-102A served the bulk of its career with the Hawaii Air Guard at Hickam AFB, where it forms this prominent display. (Alec Fushi via Terry Love)

During the 1970s, this colorfully-marked Deuce was one of two that accented the grounds at Lackland AFB, Texas. Markings of the 525th FIS were applied over those previously worn by the 37th FIS, to which this aircraft was assigned. Unusual are the commander's stripes applied vertically. (W.H. Strandberg, Jr.)

Still wearing its Wisconsin ANG markings, this weary Deuce was one of two stored at the New England Air Museum. (Lionel Paul)

During 1985, serial number 56-1140 was faithfully rebuilt and painted to represent 55-3431 of the 431st FIS. When it was presented to the McClellan AFB Museum on 13 January 1986, it was dedicated to Lt. Col. McCabe of the 431st, who was killed in a crash during 1961. (Bill Parrish via Lionel Paul)

After it was written off in 1971 following an accident in Iceland, this 57th FIS Deuce awaits restoration prior to its placement at the Air Forces Iceland Headquarters. (USAF)

Wearing a miniature 317th FIS emblem on the nose and a 21st Composite Wing badge on the tail, this F-102A went to the Alaska Transportation Museum in 1969 when Deuces were pulled out of Alaska. (Remington via Marty Isham)

This F-102A served nine units before ending its service life as a permanent display at Travis AFB. (B. Curry via Marty Isham)

transported from the MASDC beginning in 1976. Often, only the F-102's wings were used to simulate specific shapes of adversary aircraft during special ordnance evaluations. A few airframes survived the tests, some of which, according to Navy officials, "Were acquired by a private collector under less than favorable circumstances." All told, more than one-third of the F-102s produced had their careers extended beyond the era in which they served as sentinels of the skies.

F-102s Acquired by the New Mexico Institute of Mining and Technology

54-1354	55-3460	56-1041	56-1229
54-1388	55-3463	56-1097	56-1306
55-3381	55-4051	56-1102	56-1310
55-3387	56-0989	56-1138	56-1412
55-3436	56-0992	56-1187	56-1428
55-3448	56-0994	56-1188	56-1465
55-3451	56-1033	56-1203	56-1482
			56-2324

When it was moved indoors at the SAC Museum at Offutt AFB, this F-102A was repainted in the scheme it wore while assigned to the 496th FIS in Europe during the early 1960s. The unit's tail colors are repeated on the commander's stripes and the flash on the drop tank. (Jeff Kolln)

Exhibited at Minot AFB, North Dakota, this Deuce was the first placed in service with the Minnesota Air National Guard. (Mark Morgan via Marty Isham)

Originally displayed at Offutt AFB, Nebraska, in markings similar to those used by the Wisconsin Air Guard, this Deuce was later moved indoors and given a completely new scheme. (J. Vadas via Marty Isham)

Named in honor of Col. Patrick E. O'Grady, whose title appears below the cockpit, this F-102A carries an Irish theme in keeping with the colonel's heritage, down to the green shamrock behind the cockpit. Having the digits "76" within its serial number, the Deuce was an obvious candidate for Bicentennial markings during 1976. The pitot tube was painted red, white, and blue, and a special Bicentennial emblem was applied to the nose. (Marty Isham)

Following its assignment to the ADWC at Tyndall AFB, this F-102A became an exhibit at nearby Panama City. It is seen here in 1972 with ADWC markings and the legend "El Pollo Azul" on its nose. (Author)

F-102 Survivors

Serial No.	Type	Location	Remarks
53-1785	YF-102		
53-1787	YF102A	Friends of Jackson Barracks Military Museum, New Orleans, LA	
53-1788	YF-102A	McEntire ANGB, SC/Ft. Walton, FL	
53-1797	F-102A	Lackland AFB, TX	
53-1799	F-102A	Eglin AFB, TX	since removed
53-1801	F-102A	Joe Foss Field, Sioux Falls, SD	
53-1804	F-102A	Fresno Air Terminal, CA	
53-1816	F-102A	Gowen Field, Boise ANGB, ID	
53-1817	F-102A	Lackland AFB, TX	
54-1351	TF-102A	Chanute AFB, IL/Selfridge Military Air Museum, MI	
54-1353	TF-102A	Milestone of Flight Museum, Lancaster, CA	storage for Edwards AFB
54-1366	TF-102A	Pima Air and Space Museum, AZ	
54-1373	F-102A	Hickam AFB, HI	
54-1405	F-102A	New England Air Museum, CT	destroyed by tornado
54-3366	F-102A	Burlington ANGB, VT	
56-0984	F-102A	Lowry AFB, CO/Wings Over the Rockies Aviation & Space Museum, Denver, CO	
56-0985	F-102A	McEntire ANGB, SC	

Number 858 was later relocated and its tail markings significantly altered. (Marty Isham)

Complete with 146th FIS markings and a pilot mannequin in the cockpit, this Deuce stands guard along a roadway at Pennsylvania's Greater Pittsburgh Airport. (Mark Morgan via Marty Isham)

This Tub is one of two Delta Daggers among a number of aircraft exhibits at Wisconsin's Volk Field. Both aircraft once served the 176th FIS. (Author)

The TF-102A's companion at Volk Field. A white helmet covered with red hearts is visible in the cockpit. (Author)

56-0986	F-102A	Florida Military Aviation Museum, Clearwater, FL	
56-0989	F-102A	Lowry AFB, CO	
56-0995	F-102A	Santa Maria Park, CA	
56-0997	F-102A	Sheppard AFB, TX	
56-1017	F-102A	South Dakota Air & Space Museum, Ellsworth AFB, SD	
56-1102	F-102A	Great Falls Airport, MT	
56-1105	F-102A	Great Falls ANGB, MT	
56-1106	F-102A	Tanagra AB, Greece/Ellsworth AFB, SD	
56-1109	F-102A	Peterson Air & Space Museum, Peterson AFB, CO	
56-1114	F-102A	March Field Museum, CA	
56-1115	F-102A	Fairchild AFB, WA	
56-1116	F-102A	Helena College of Technology, MT	civil reg. N8970
56-1125	F-102A	Eifel Museum, Germany	
56-1134	F-102A	Tucson ANGB, AZ	
56-1140	F-102A	McClellan Aviation Museum, CA	
56-1151	F-102A	Lackland AFB, TX	
56-1204	F-102A	Chanute AFB, IL	
56-1219	F-102A	Syracuse, NY/Minnesota Air Guard Museum, Minneapolis, MN	as 56-1476
56-1221	F-102A	New England Air Museum, CT	
56-1247	F-102A	Travis AFB, CA	
56-1252	F-102A	Ellington AFB, TX	
56-1264	F-102A	Bradley Airport, CT	

First mounted for display at Air Forces Iceland Headquarters in 1971 following an accident at NAS Keflavik, this F-102A wore three completely different schemes before it succumbed to high winds and was destroyed. The Deuce was originally presented as it appeared while assigned to the 57th FIS. (Baldur Sveinsson)

To celebrate America's Bicentennial, Keflavik's F-102 gate guard shed its 57th livery to don camouflage and Bicentennial markings. (Baldur Sveinsson)

Keflavik's F-102A as it appeared in 1978 in its final scheme. (Baldur Sveinsson)

56-1266	F-102A	Ernest Harmon Airport, Newfoundland
56-1268	F-102A	Kelly AFB, TX
56-1273	F-102A	Volk Field, WI
56-1274	F-102A	Elmendorf AFB, AK
56-1282	F-102A	Museum of Transportation & Industry, Wasilla, AK
56-1357	F-102A	Jacksonville Airport, FL
56-1365	F-102A	Syracuse Park, NY
56-1368	F-102A	Oregon Museum of Science and Industry, Portland, OR
56-1378	F-102A	NAS Keflavik, Iceland
56-1393	F-102A	Pima Air & Space Museum, AZ
56-1413	F-102A	Planes of Fame, Chino, CA
56-1415	F-102A	Greater Pittsburgh Airport, PA
56-1431	F-102A	NAS China Lake, CA
56-1432	F-102A	Camp Robinson, AR
56-1502	F-102A	Fargo ANGB, ND
56-1505	F-102A	Duluth Airport, MN/Minot AFB, ND
56-2317	TF-102A	Yankee Air Force, Belleville, MI
56-2333	TF-102A	Eglin AFB, FL
56-2337	TF-102A	Lackland AFB, TX
56-2346	TF-102A	Ft. Indiantown Gap, PA
56-2352	TF-102A	Kelly AFB, TX
56-2353	TF-102A	Volk Field, TX
56-2364	TF-102A	Sacramento Airport, CA
57-0775	F-102A	Clovis Park, CA
57-0777	F-102A	NAS China Lake, CA
57-0788	F-102A	West Hampton Beach ANGB, NY
57-0812	F-102A	Kirtland AFB, NM
57-0817	F-102A	Jacksonville, FL
57-0826	F-102A	Sheppard AFB, TX
57-0833	F-102A	Hill AFB, UT
57-0858	F-102A	Panama City, FL
57-0905	F-102A	Ontario Airport, CA
57-0906	F-102A	Confederate AF, Harlingen, TX/Robins AFB, GA

Deuce Gallery

F-102A of the 326th FIS at Richards-Gebaur AFB. The unpainted radome matched the trim color of the aircraft. (Terry Love)

TF-102A of the California Air Guard's 194th FIS at Fresno Airport in October 1967. The squadron's parent unit, the 144th Fighter Group, was identified on the aircraft's drop tank. (Stephen Miller)

Assigned to the 482nd FIS, this Deuce is seen at Andrews AFB, Maryland, during May 1965. (Stephen Miller)

A Tub of the 57th FIS in 1969. The aircraft's high-visibility fluorescent red-orange trim color, which was susceptible to weathering, was replaced in favor of Insignia Red, as seen on the Deuce to its right. A large Air Force Outstanding Unit Award was applied to the forward fuselage. (Ian Macpherson via Stephen Miller)

A Deuce of the 71st FIS in 1960. (John Anderson via Stephen Miller)

A TF-102A of the Idaho ANG at Boise during April 1966. In Air Guard tradition, the parent command (124th FIG) is noted on the drop tank. The radome and anti-glare panel have a thin white outline. (Paul Stevens via Stephen Miller)

A TF-102A of the Europe-based 496th FIS. (Jack M. Friell)

Ground crewmen contend with the weather at Thule AB, Greenland, in 1962 to prepare a Tub of the 332nd FIS for a mission. (Jerry Geer via Nick Williams)

A JF-102A undergoes cold weather tests at Ladd AFB, Alaska, during February 1956. (Larry Davis)

A freshly camouflaged F-102A of the Idaho ANG at Boise in June 1973. (Udo Weisse via Stephen Miller)

Idaho's 190th FIS was unique in that it used a lime-yellow color to highlight its F-102s. This example carries an electronics pod on a wing pylon. (Baldur Sveinsson Collection)

A Tub of Arizona's 152nd FIS at Tucson in October 1967. (Stephen Miller)

The Group designation of this Idaho ANG Deuce is plainly indicated on the drop tank. (Hugh Muir)

A Deuce of the 190th FIS in May 1967. The trim color was applied to the pitot tube and the air inlet section was outlined with red. (Robert Burgess via Stephen Miller)

Painted with a shade of light gray somewhat lighter than the standard ADC Gray, this TF-102A wears the Air Training Command badge on its tail fin. This was the first-built trainer variant. (Jack M. Friell)

An immaculate F-102A of the 194th FIS. (Terry Love)

This TF-102A rests in storage after having served with the Wisconsin Air Guard. The Tub's radome and anti-glare panel were trimmed with a thin yellow and black, while a black and yellow band with an intricate weave pattern borders the air intake. A "TWA" sticker is visible near the top of the tail fin. (Frank MacSorley via Stephen Miller)

After serving the 317th FIS and Wisconsin's 176th FIS, this TF-102A was relegated to desert storage during 1974. (Ted Paskowski via Terry Love)

Minus its exhaust section, this badly weathered Deuce was pulled out of storage in 1979 for use in the drone program, where it was destroyed in 1981. (Werner Hartman via Terry Love)

An F-102A of the 157th FIS at Eastover, South Carolina, in September 1969. (Nelson Hare via Stephen Miller)

After trying various marking schemes, the Pennsylvania Air Guard settled on this attractive pattern for its Deuces. (Larry Davis Collection)

A TF-102A of Montana's 186th FIS at Great Falls in April 1972. (Robert J. Pickett)

A Deuce of the Montana ANG with the short-lived orange high-visibility markings in 1972. (L.B. Sides)

This early model Deuce was eventually taken from its desert roost to become a permanent display in South Dakota. (Terry Love)

Formerly of the 122nd FIS, this F-102A's openings and panel lines were sealed for storage. (Terry Love)

Wearing a unit emblem on its nose, an F-102A of the 199th FIS takes off from Hickam AFB, Hawaii, in January 1968. (Nick Williams)

Shrouds were commonly used in hot climates to protect the F-102's cockpit, radar, and fire control system. This Deuce of the 159th FIS is seen in 1968. (Nick Williams)

An F-102A of the 118th FIS at Bradley Field, Connecticut, on 30 May 1970. (Nick Williams)

TF-102A of the 196th FIS at Fresno, California, in October 1967. This Tub was later passed to the Greek Air Force. (Stephen Miller)

This Delta Dagger served with a number of units before it underwent conversion to a drone. It is seen here while assigned to the California Air Guard. (Author)

Seen in the unmistakable tail markings of the 57th FIS, this Deuce also wore the special black knight and bear markings on its drop tanks. (Candid Aero-Files)

A Deuce of the 327th FIS is prepared for a mission from Thule AB, Greenland, in sub-zero temperatures during February 1959. (Budd Butcher)

Besides the standard unit markings, liberal amounts of red paint were used to highlight panel lines on the forward fuselages of 176th FIS F-102s. (L.B. Sides)

The 37th FIS used a patriotic theme in marking its aircraft. The red area behind the cockpit of this Deuce, seen in 1958, also appeared in blue. (Ron Picciani via Marty Isham)

An F-102A of the 61st FIS at Scott AFB, Illinois, in May 1960. The aircraft was destroyed in a crash in 1964. (Robert Burgess via Marty Isham)

Though commander's stripes varied in color, they were usually applied diagonally somewhere on the F-102's fuselage. This example hailed from the 460th FIS in 1962. (W. Jefferies via Marty Isham)

The 326th FIS participated in three of the first five William Tell competitions, taking top honors in the 1958 meet. This "Skywolves" Deuce wears a single stripe across the top of its fuselage. (Author)

This Deuce of the 57th FIS wears the unit's early pattern of blue and white checks on its rudder. (Nick Williams)

An F-102A of the ADWC at Tyndall AFB in 1972. (L.B. Sides)

This Deuce of the 326th FIS wears the squadron "Skywolves" insignia it adopted in 1961. (L.B. Sides)

A TF-102A of the 318th FIS in 1957, the year that the squadron exchanged its F-86s for F-102s. (D. Henderson via Marty Isham)

A Deuce of the 57th FIS "Black Knights." (L.B. Sides)

A 57th FIS F-102A with early rudder markings in 1969. (Jack M. Friell)

This Deuce ended its extensive service with Hawaii's 199th FIS. It is seen here at Hickam AFB in September 1963. (Larry Davis Collection)

Named the "City of Erie," a 146th FIS Deuce is readied for a mission. (Hugh Muir via Terry Love)

Framed by Iceland's colorful scenery, a Tub of the 57th FIS taxis at NAS Keflavik. (Baldur Sveinsson)

The red markings generously applied to this F-102A of the 11th FIS were a carryover from its previous assignment to the 37th FIS. The aircraft is seen here in August 1959. (Jerry Geer)

The wing commander's aircraft of the 4780th ADW at Perrin AFB, Texas, during late 1964. Prior to duty with the 4780th, this Deuce was one of two JEF-102As attached to the 526th FIS in Germany for trials with new data link weapons and control systems. (Author)

The 11th FIS commander's aircraft in 1957. (David Menard)

A Deuce of Hawaii's 199th FIS in January 1968. (Nick Williams)

This TF-102A of the 59th FIS, seen at Andrews AFB during June 1965, wears blue trim to accent its red panels. (Dr. J.C. Handelman via Marty Isham)

The 27th FIS changed its markings when it received F-102s, but retained the original falcon in the center. This Deuce appears at Griffiss AFB in 1958. (Gene Bellveau)

Various tail fin tip colors were used by Vermont's 134th FIS to identify flights within the squadron. This 134th F-102A is seen at Pease AFB during August 1969. (Jack Morris via Marty Isham)

This 325th FIS Deuce wore its commander's stripes on the forward fuselage. (Terry Love)

This 176th FIS Deuce, seen at an air show in Milwaukee, Wisconsin, during August 1967, was later turned over to the Turkish Air Force. (R.M. Hill)

An F-102A of the 182nd FIS wears the 149th Fighter Group badge, along with the title "149th FTR.GP. KELLY AFB" on the drop tank. (Jerry Geer Collection)

An F-102A of the 86th FIS at Youngstown, Ohio, in May 1959 is equipped with drop tanks that featured a two-color horizontal line pierced by three lightning bolts. (Marty Isham Collection)

A TF-102A of the 86th FIS in May 1959. (Marty Isham Collection)

Yellow stripes were superimposed over the red tail fin color of this Deuce of the 61st FIS. (Roger Besecker)

An F-102A (s/n 56-1211) of New York's 102nd FIS. (Author)

This fully upgraded TF-102A served the North Dakota and South Carolina Air Guards before it was placed in storage. (Author)

Seen here in 1961, this Deuce was assigned to the 325th FIS, which was based at Truax Field, Wisconsin, from early 1957 to mid-1966. The unit's tail markings comprised white and blue stripes divided by a red chevron, above which was a 30th Air Division badge. (Leo Kohn)

The red and yellow commander's stripes on this 18th FIS F-102A were a deviation from the standard fuselage location. (Marty Isham Collection)

Wrap-around commander's stripes were among the markings that adorned this Deuce of the 327th FIS at George AFB. The aircraft was assigned to the squadron's "Spade Flight." The name "The Challenger" on the nose landing gear door was accompanied by a depiction of a knight riding a horse. (USAF)

A Deuce wearing the distinctive emblem of the 460th FIS lands at Paine Field, Washington. (Marty Isham)

A TF-102A of the California Air Guard at Elmendorf AFB, Alaska, in August 1968. Two years later, it was transferred under the MAP to Greece. (Lionel Paul)

An F-102A of the 196th FIS. (L.B. Sides)

A Delta Dagger of California's 194th FIS at Sheppard AFB, Texas, during July 1967. (Merle Olmsted)

A tiny squadron emblem appeared on the tip of the fairing that extended beyond the exhaust of this 57th FIS Deuce. (Author)

The home base of this 4780th ADW F-102A was announced on the external fuel tank markings. Seen in May 1963, the aircraft carried the emblem of its parent command, the 73rd AD, on the tail fin. (Merle Olmsted)

After this Deuce crashed in 1958, it retained its 329th FIS tail markings when it became a maintenance trainer at Sheppard AFB. "Adelanto" is the California city near George AFB. (Author)

Trailing its drag chute, an F-102A of the 57th FIS taxis at NAS Keflavik in 1966. (Baldur Sveinsson)

Clearly identified by its markings unique to the Pennsylvania Air Guard, and wearing the name "City of Ambridge," this Deuce entered the drone program during February 1975. (L.B. Sides)

A Deuce of the 176th FIS. (USAF)

Following service with the South Carolina Air Guard, this F-102A became a permanent exhibit in the state. (Author)

This F-102A wears the original insignia of the 146th FIS. (Lionel Paul)

A Deuce of the 178th FIS in December 1968. (AAHS/Jerry Geer)

The small yellow lightning bolts worn on the tail fins of 176th FIS F-102s paled in comparison to the bold red markings worn later. (Author)

Having just been transferred to the Louisiana Air Guard in September 1960, this Deuce still wears the markings of the 323rd FIS. It was destroyed in a mid-air collision with another F-102 one year later. (LA ANG)

An American Airlines emblem replaced the standard ANG decal normally worn on the tail of 176th FIS F-102s. Since another of the squadron's Deuces sported a TWA logo, there may have been some rivalry between Deuce pilots who flew for commercial airlines on a full time basis. (David McLaren Collection)

F-102s were among a wide variety of aircraft flown by New York's 102nd FIS. They replaced KC-97L tankers and were themselves replaced by rescue versions of the C-130 Hercules and H-3 helicopter. Most 102nd F-102s were camouflaged shortly after their arrival to the squadron. (Ron Montgomery via Leo Kohn)

Deuces served with New York's Air Guard for a two and a half year period. This upgraded F-102A makes a low-level pass near its home base at Suffolk County Airport, Westhampton Beach. (Ron Montgomery via Leo Kohn)

A TF-102A of the 159th FIS during June 1971. (Jerry Geer)

This Deuce served with both the 111th and 182nd FIS of the Texas Air Guard. Lettering on the fuselage would later read "U.S. Air Force." (Ron Montgomery via Leo Kohn)

Left: An F-102A of the 157th FIS, based at McEntire ANGB in 1968. (USAF)

Connecticut's 118th FIS was called "The Flying Yankees," which was advertised on the aircrafts' fuel tanks. The squadron insignia was worn forward of the cockpit, and the aircraft number, carried on the tail fin tip, was repeated on the aft fuselage. (Ron Montgomery via Leo Kohn)

A Tub of the 102nd FIS at Long Island in 1973. (Lionel Paul)

A Deuce of the 47th FIS at Niagara Falls Airport in early 1959. Tail markings comprised a 3 and 4 of dice superimposed over a yellow "4" and black spade on a yellow band, edged in black. (Fred T. Guthrie via Marty Isham)

An F-102A of the 325th FIS, based at Truax Field, Wisconsin, in 1963. (Author)

A Deuce of the 11th FIS, based at Duluth Airport, Minnesota. This aircraft was destroyed in a crash during 1968. (TX ANG)

An early model Deuce of the 327th FIS lifts from the runway at George AFB, California. (Pima Air & Space Museum)

F-102s of the 48th FIS sported these dazzling tail markings throughout most of the time the unit operated Deuces. This F-102A is seen at Congaree AB, South Carolina, in September 1959. (Author)

An F-102A of the 326th FIS wearing the squadron's original insignia and three commander's stripes at Peterson Field in 1959. (R. McCarthy via Marty Isham)

One of the few silver-painted TF-102As, this Tub is seen during its service with the 4780th ADW at Perrin AFB, Texas. (Marty Isham Collection)

The Deuce looked imposing, yet graceful from any angle. This early model wears the ARDC badge on its nose during tests at Edwards AFB in 1957. (Jerry Geer Collection)

An F-102A of the 326th FIS in pristine condition in 1961. (USAF)

F-102 Emblems

2nd FIS - ADC

4th FIS - PACAF

5th FIS - ADC

Possibly 16th FIS

11th FIS - ADC

16th FIS - PACAF

18th FIS - ADC

27th FIS - ADC

31st FIS - ADC/AAC

32nd FIS - USAFE

37th FIS - ADC

40th FIS - PACAF

Below: 40th FIS - (FEAF)

40th FIS - PACAF

59th FIS - ADC

57th FIS - ADC

47th FIS - ADC

59th FIS - ADC

Left: 48th FIS - ADC

Right: 64th FIS - ADC/ PACAF

61st FIS - ADC

68th FIS - PACAF

71st FIS - ADC

76th FIS - ADC

Left: 82nd FIS -
ADC/PACAF

86th FIS - ADC

87th FIS - ADC

95th FIS - ADC

317th FIS - ADC/AAC

317th FIS - ADC/AAC

325th FIS - ADC

318th FIS - ADC

323rd FIS - ADC

326th FIS (original) - ADC

326th FIS - ADC

327th FIS - ADC

329th FIS - ADC

331st FIS - ADC

332nd FIS - ADC

431st FIS - USAFE

438th FIS - ADC

456th FIS - ADC

460th FIS - ADC

482nd FIS - ADC

496th FIS - ADC

497th FIS - ADC

498th FIS - ADC

498th FIS - ADC

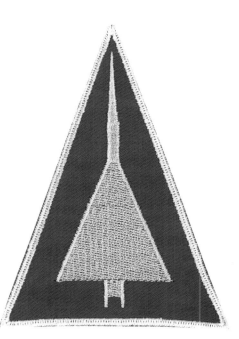

509th FIS - PACAF

525th FIS - USAFE

526th FIS - USAFE

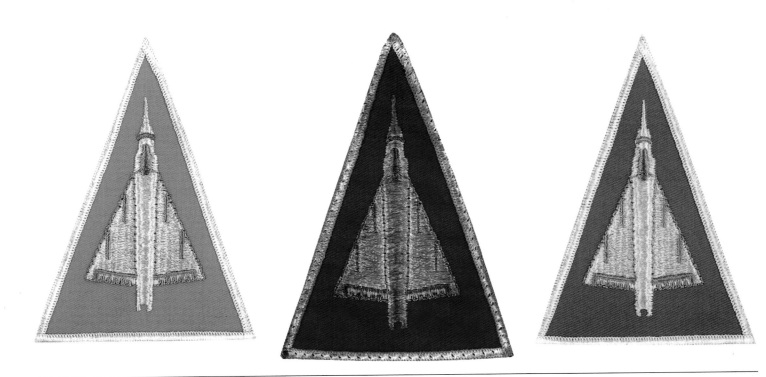

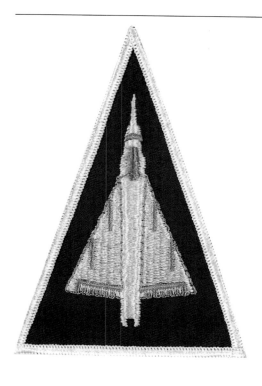

ADC Qualification Badge

ADC Qualification Badge

PACAF

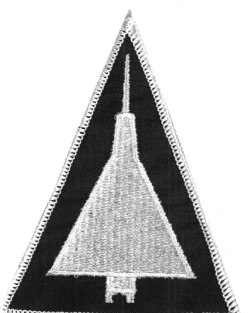

325th FIS

32nd FIS

59th FIS

AFFTC

ADWC

4780th ADW

3556th FTS

4781st CCTS

4782nd CCTS

3558th FTS

Perrin AFB, Denison, TX

IWS

ADC "A" Award

525th FIS - USAFE

475th TS

4750th TS

82nd ATS, Tyndall AFB

82nd ATS, Det. 1 Holloman AFB

RCA - Drone Program

1959 William Tell Team Emblem

176th FIS - WI ANG

176th FIS Aircrew

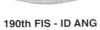

190th FIS - ID ANG

Left: 176th FIS Groundcrew

199th FIS - HI ANG

163rd FIG - CA ANG

194th FIS - CA ANG

196th FIS - CA ANG

132nd FIS - ME ANG

116th FIS - WA ANG

102nd FIS - NY ANG

MT ANG

178th FIS - ND ANG

123rd FIS - OR ANG

111th FIS - TX ANG

Appendix A
F-102 Production

DESIGNATION	MODEL	USAF SERIAL NO.	UNITS BUILT
YF-102	8-80	52-7994/7995	2
YF-102	8-82	53-1779/1786	8
YF-102A-17-CO	8-90	53-1787/1790	4
F-102A-5-CO	8-10	53-1791/1794	4
F-102A-IO-CO	8-10	53-1795/1797	3
F-102A-15-CO	8-10	53-1798/1802	5
F-102A-20-CO	8-10	53-1803/1811	9
F-102A-25-CO	8-10	53-1812/1818	7
F-102A-30-CO	8-10	54-1371/1383	13
F-102A-35-CO	8-10	54-1384/1400	17
F-102A-40-CO	8-10	54-1401/1407	7
F-102A-41-CO	8-10	55-3357/3379	23
F-102A-45-CO	8-10	55-3380/3426	47
F-102A-50-CO	8-10	55-3427/3464	38
F-102A-51-CO	8-10	56-0957/0972	16
F-102A-55-CO	8-10	56-0973/1044	72
F-102A-60-CO	8-10	56-1045/1136	92
F-102A 65-CO	8-10	56-1137/1233	97
F-102A-70-CO	8-10	56-1234/1274	41
F-102A-75-CO	8-10	56-1275/1316	42
F-102A-65-CO	8-10	56-1317/1320	4
F-102A-70-CO	8-10	56-1321/1331	11
F-102A-75-CO	8-10	56-1332/1429	98
F-102A-80-CO	8-10	56-1430/1518	89
F-102A-90-CO	8-10	57-0770/0855	76
F-102A-95-CO	8-10	57-0856/0909	54
TF-102A-5-CO	8-12	54-1351/1354	4
TF-102A-10-CO	8-12	54-1355/1359	5
TF-102A-35-CO	8-12	54-1360	1
TF-102A-15-CO	8-12	54-1361/1365	5
TF-102A-20-CO	8-12	54-1366/1368	3
TF-102A-25-CO	8-12	54-1369/1370	2
TF-102A-26-CO	8-12	55-4032/4034	3
TF-102A-30-CO	8-12	55-4035/4042	8
TF-102A-35-CO	8-12	55-4043/4050	8
TF-102A-36-CO	8-12	55-4051/4056	6
TF-102A-37-CO	8-12	55-4057/4059	3
TF-102A-35-CO	8-12	56-2317/2323	7
TF-102A-40-CO	8-12	56-2324/2335	12
TF-102A-41-CO	8-12	56-2336/2353	18
TF-102A-45-CO	8-12	56-2354/2379	26

Appendix B
Acronyms and Abbreviations

AAC Alaskan Air Command
AB Air Base
ABW Air Base Wing
AD Air Defense/Air Division
ADC Air Defense Command/Aerospace Defense Command
ADG Air Defense Group
ADS Air Defense Sector
ADTC Armament Development Test Center
ADW Air Defense Wing
ADWC Air Defense Weapons Center
AEW Airborne Early Warning
AF Air Force
AFAC Air Force Armament Center
AFB Air Force Base
AFCARC Air Force Cambridge Air Research Center (Hanscom AFB)
AFCS Automatic Flight Control System
AFFTC Air Force Flight Test Center
AFI Air Forces Iceland
AFLC Air Force Logistics Command
AFMDC Air Force Missile Development Center
AFOG Air Force Operations Group
AFS Air Force Station
AFSC Air Force Systems Command
AFSWC Air Force Special Weapons Center
AIM Air Intercept Missile
AMA Air Material Area
AMARC Aircraft Maintenance And Regeneration Center
AMC Air Material Command
AMFEA Air Material Force European Area
ANG Air National Guard
ANGB Air National Guard Base/Bureau
APGC Air Proving Ground Center
ARDC Air Research and Development Center/Command
ASD Aeronautical Systems Division (USAF)
ATC Air Training Command
BMEWS Ballistic Missile Early Warning Site
CADF Central Air Defense Force
CAMRON Consolidated Aircraft Maintenance Squadron (also CAMS)
CCTS Combat Crew Training Squadron
CONAC Continental Air Command

CONAD Continental Air Defense Command
DEW Distant Early Warning
DoD Department of Defense
EADF Eastern Air Defense Force
FAA Federal Aviation Administration
FEAF Far East Air Forces
FFAR Folding Fin Aerial Rocket
FG Fighter Group
FIG Fighter Interceptor Group
FIS Fighter Interceptor Squadron
FIW Fighter Interceptor Wing
FTW Flying Training Wing
GAR Guided Aerial Rocket
HADC Holloman Air Development Center
ICBM Inter-Continental Ballistic Missile
IP Instructor Pilot
IRAN Inspection and Repair As Necessary
IWS Interceptor Weapons School
MAP Military Assistance Program
MASDC Military Aircraft Storage and Disposition Center
MDAP Mutual Defense Assistance Program
NACA National Advisory Committee for Aeronautics
NAS Naval Air Station
NASA National Aeronautics and Space Administration
NATO North Atlantic Treaty Organization
NEAC Northeast Air Command
NMIMT New Mexico Institute of Mining and Technology
NORAD North American Air Defense Command
NWC Naval Weapons Center
PACAF Pacific Air Forces
SAC Strategic Air Command
SAGE Semi-Automatic Ground Environment
SEA Southeast Asia
SHAPE Supreme Headquarters Allied Powers Europe
SVN South Vietnam
TAC Tactical Air Command
TFS Tactical Fighter Squadron
TG Test Group
TS Test Squadron
USAFE United States Air Forces in Europe
WADC Wright Air Development Center
WADF Western Air Defense Force
WSEM Weapons System Evaluation Missile

Appendix C
F-102 Individual Aircraft Record

YF-102
52-7994
52-7995 USAF Museum/Fairborn, Ohio display; later scrapped
53-1779
53-1780
53-1781
53-1782
53-1783
53-1784
53-1785 NASA; AFFTC
53-1786

YF-102A
53-1787 AFFTC; military museum, New Orleans, LA
53-1788 SC ANG; McEntire ANGB display; Ft. Walton, FL display
53-1789

F-102A
53-1790
53-1791 ADC; AMC; 327th FIS; 16th FIS; 509th FIS; accident/ destroyed 9-7-67
53-1792 460th FIS; AMC; 16th FIS; 68th FIS; WA ANG; MASDC Dec. 1969
53-1793 18th FIS; 37th FIS; 460th FIS; 16th FIS; 509th FIS
53-1794
53-1795 37th FIS
53-1796
53-1797 Lackland AFB display
53-1798 509th FIS
53-1799 37th FIS; 327th FIS; weapons test; Eglin AFB display
53-1800
53-1801 37th FIS; 496th FIS; ID ANG; Sioux Falls Airport display as 56-1114
53-1802 JF-102A ARDC; 16th FIS; 317th FIS; 509th FIS; ID ANG; MASDC May 1971
53-1803 18th FIS; 37th FIS; 460th FIS; 68th FIS; CA ANG; MASDC Jan. 1970
53-1804 37th FIS; 82nd FIS; 18th FIS; 460th FIS; 68th FIS; CA ANG; Fresno Air Terminal, CA display
53-1805 68th FIS; 460th FIS; CA ANG; MASDC Dec. 1969
53-1806 YF-102C; 37th FIS; 525th FIS
53-1807 ARDC; 496th FIS; ID ANG; MASDC July 1971
53-1808 ARDC; 460th FIS; 16th FIS; 509th FIS
53-1809 37th FIS; 496th FIS; ID ANG; MASDC March 1971
53-1810 37th FIS; 496th FIS; 497th FIS; ID ANG; CA ANG; MASDC Nov. 1971
53-1811 37th FIS; 496th FIS; ID ANG; MASDC Dec. 1971
53-1812 37th FIS; 3211th Interceptor Test Group; Gowen Field, Boise, ID display
53-1813 460th FIS; 16th FIS; 509th FIS
53-1814 37th FIS; 460th FIS; 68th FIS; CA ANG; Turkey
53-1815 82nd FIS; 460th FIS; 509th FIS; MT ANG; Turkey
53-1816 37th FIS; 496th FIS; ID ANG; Gowen Field, Boise, ID display
53-1817 37th FIS; AFLC; Lackland AFB display
53-1818 37th FIS; 496th FIS; crash/destroyed 7-20-60
54-1371 AFCARC/ARDC; 460th FIS; 509th FIS; HI ANG
54-1372 AFCARC/ARDC; 337th CAMS; 460th FIS; 509th FIS; HI ANG; MASDC Dec. 1976
54-1373 82nd FIS; AFCARC/ARDC; HI ANG; Hickam AFB display
54-1374 NASA; MASDC May 1959
54-1375 37th FIS; 82nd FIS; 496th FIS; destroyed by fire 11-22-61
54-1376 APGC; 37th FIS; 82nd FIS; 496th FIS; 32nd FIS; 497th FIS; ID ANG; MASDC March 1970
54-1377 82nd FIS; 460th FIS; 68th FIS; CA ANG; Turkey
54-1378 37th FIS; 496th FIS; crash/destroyed 8-22-60
54-1379 327th FIS; 438th FIS; 37th FIS; 460th FIS; 68th FIS; 431st FIS; CA ANG; Turkey
54-1380 37th FIS; 82nd FIS; CA ANG; PA ANG; Turkey
54-1381 327th FIS
54-1382 37th FTS; 460th FIS; 68th FIS; 431st FIS; CA ANG; Turkey
54-1383 327th FIS; 438th FIS; 37th FIS; 68th FIS; CA ANG; Turkey
54-1384 37th FIS; 327th FIS; 460th FIS; 68th FIS; CA ANG; Turkey
54-1385 37th FIS; 327th FIS; 82nd FIS; 496th FIS; ID ANG; drone
54-1386 37th FIS; 327th FIS; 18th FIS; 460th FIS; CA ANG; Turkey
54-1387 68th FIS; 460th FIS; 496th FIS; 509th FIS; CA ANG; MASDC Jan. 197
54-1388 327th FIS; 460th FIS; 16th FIS; WI ANG; CA ANG; NMIMT
54-1389 327th FIS; 37th FIS; 496th FIS; ID ANG; MT ANG; crash/destroyed 12-30-71
54-1390 ARDC
54-1391 37th FIS; 327th FIS; 68th FIS; 18th FIS; 460th FIS; CA ANG; MASDC March 1971
54-1392 37th FIS; 327th FTS; 496th FIS; ID ANG; ND ANG; MASDC Dec. 1969; Turkey

54-1393 37th FIS; 327th FIS; 496th FIS; ID ANG; drone
54-1394 37th FIS; 327th FIS; 496th FIS; 32nd FIS; ID ANG; crash/destroyed 4-30-64
54-1395 37th FIS; 327th FTS; 496th FIS; 431st FIS; ID ANG; drone
54-1396 327th FIS; 68th FIS; 82nd FIS; 460th FIS
54-1397 37th FIS; 327th FIS; 496th FIS; 526th FIS; ID ANG; San Diego Airport display
54-1398 37th FIS; landing gear tests; J85 tests
54-1399 37th FIS; 327th FIS; 496th FIS; 32nd FIS; MT ANG; ID ANG
54-1400 37th FIS; 327th FIS; 496th FIS; MT ANG; ID ANG; MASDC March 1970
54-1401
54-1402 327th FIS; 37th FIS; 496th FIS; 59th FIS; 4780th ADW; WI ANG
54-1403 327th FIS; 37th FIS; MT ANG; Turkey
54-1404 327th FIS; 37th FIS; 525th FIS; 496th FIS; crash/destroyed 1-24-64
54-1405 327th FIS; 14th CAMS; 86th FIS; 496th FIS; 32nd FIS; 59th FIS; 4780th ADW; WI ANG; Turkey; SAC Museum
54-1406 37th FIS; 327th FIS; 496th FIS; 32nd FIS; 497th FIS; ID ANG; drone
54-1407 37th FIS; 496th FIS; ID ANG; drone
55-3357 11th FIS; 86th FIS; HI ANG
55-3358 11th FIS; 2nd FIS
55-3359 11th FIS; 327th FIS; 86th FIS; 16th FIS; HI ANG
55-3360 11th FIS; 2nd FIS; 16th FTS; HI ANG
55-3361 11th FIS
55-3362 11th FIS, 2nd FIS; 327th FIS; 496th FIS; 64th FIS; 16th FIS; 509th FTS; crash/destroyed SVN 4-2-67
55-3363 11th FIS; 2nd FIS; 509th FIS
55-3364 11th FIS; 2nd FIS; HI ANG
55-3365 11th FIS
55-3366 11th FIS; 2nd FIS; 16th FIS; HI ANG; Hickam AFB display
55-3367 11th FIS; 86th FIS; 16th FIS; 509th FIS
55-3368 11th FIS; 2nd FIS; 16th FIS; HI ANG
55-3369 11th FIS; 2nd FIS; 16th FIS; 509th FIS
55-3370 327th FIS; 86th FIS; 16th FIS; 509th FIS; HI ANG
55-3371 327th FIS; 86th FIS; 16th FIS; 509th FIS; destroyed ground attack SVN 7-1-65
55-3372 327th FIS; 86th FIS; 16th FIS; crash/destroyed 1-28-63
55-3373 318th FIS; 325th CAMS; MT ANG; 509th FIS; shot down SVN 12-15-65
55-3374 11th FIS; 2nd FIS; 509th FIS; HI ANG
55-3375 11th FIS; 86th FIS; 509th FIS
55-3376 11th FIS; 86th FIS; HI ANG
55-3377 11th FIS; 2nd FIS; HI ANG
55-3378 327th FIS; 86th FIS; HI ANG; ID ANG
55-3379 327th FIS; 86th FIS; 16th FIS; 509th FIS
55-3380 327th FIS; 47th FIS; 40th FIS; CT ANG; MT ANG; Turkey; crash/destroyed 9-18-69
55-3381 323rd FIS; 318th FIS; 68th FIS; HI ANG; CA ANG; NMIMT
55-3382 327th FIS; 47th FIS; crash/destroyed 12-23-58
55-3383 327th FIS; 47th FIS; 40th FIS; 68th FIS; 431st FIS; MT ANG; Turkey
55-3384 323rd FIS; 47th FIS; CA ANG; Turkey
55-3385 CA ANG; Turkey
55-3386 327th FIS; 318th FIS; HI ANG; CA ANG; Turkey
55-3387 323rd FIS; 86th FIS; 47th FIS; 40th FIS; CA ANG; MASDC Jan. 1970; NMIMT
55-3388 323rd FIS; 86th FIS; 16th FIS
55-3389 327th FIS; 323rd FIS; 318th FIS; 327th FIS; 4780th ADW; CA ANG; Turkey
55-3390 11th FIS; 318th FIS; MT ANG; Turkey
55-3391 11th FIS; 318th FIS; NASA; 4780th ADW; CA ANG; MASDC Dec. 1969
55-3392 11th FIS; 318th FIS; 40th FIS; CT ANG; Turkey
55-3393 327th FIS; 47th FIS; 4780th ADW; 496th FIS; 32nd FIS; ND ANG; AZ ANG; MASDC Nov. 1969
55-3394 323rd FTS; 318th FIS; crash/destroyed 6-4-59
55-3395 11th FIS; 318th FIS; 40th FIS; CT ANG; Turkey
55-3396 323rd FIS; 86th FIS; 47th FIS; CA ANG; Turkey
55-3397 323rd FIS; 86th FIS; 318th FIS; 40th FIS; CA ANG; crash/destroyed 12-10-66
55-3398 11th FIS; 318th FIS; 4780th ADW; MT ANG; CA ANG
55-3399 11th FIS; 47th FIS
55-3400 327th FIS; 47th FIS; 40th FIS; MT ANG; WI ANG; Turkey
55-3401 327th FIS; 318th FIS; 4780th ADW; CA ANG; MASDC Jan. 2970
55-3402 327th FIS; 318th FIS; 40th FIS; CT ANG; MASDC May 1970
55-3403 327th FIS; 318th FIS; 40th FIS; MT ANG; Turkey
55-3404 11th FIS; 47th FIS; 4th; 16th FIS; MT ANG; CT ANC; Turkey
55-3405 327th FIS; 318th FIS; 4780th ADW; NASA; CT ANG; Turkey
55-3406 323rd FIS; 86th FTS; crash/destroyed 12-23-57
55-3407 327th FIS; 47th FIS; CA ANG; MASDC Jan. 1970
55-3408 498th FIS; 47th FIS; 40th FIS; 526th FIS; CT ANG; Turkey
55-3409 MT ANG; Turkey

55-3410 327th FIS; 318th FIS; 4780th ADW; CA ANG; Turkey
55-3411 327th FIS; 318th FIS; 496th FIS; 32nd FIS; 526th FIS; 4780th ADW; WI ANG; crash/destroyed 8-29-67
55-3412 327th FIS; 47th FIS; 40th FIS; CT ANG; Turkey
55-3413 323rd FIS; 86th FIS; 47th FIS; CA ANG
55-3414 323rd FIS; 86th FIS; 47th FIS; 40th FIS; CA ANG; MASDC Jan. 1970
55-3415 323rd FIS; 86th FIS; crash/destroyed 10-29-57
55-3416 327th FIS; 318th FIS; 40th FIS; MT ANG; Turkey
55-3417 327th FIS; 318th FIS; 4780th ADW; SD ANG; MT ANG
55-3418 323rd FIS; 86th FIS; 47th FIS; 4780th ADW; 496th FIS; 32nd FIS; WA ANG; WI ANG; MASDC Jan. 1970
55-3419 327th FIS; 47th FIS; 40th FIS; CT ANG
55-3420 327th FIS; 47th FIS; 40th FIS; 431st FIS; MT ANG; Turkey
55-3421 327th FIS; 47th FIS; 40th FIS; CA ANG; Turkey
55-3422 31st FIS; 318th FIS; 4780th ADW; crash/destroyed 9-18-63
55-3423 327th FIS; 47th FIS; 40th FIS
55-3424 323rd FIS; 68th FIS; WA ANG; crash/destroyed 6-24-69
55-3425 323rd FIS; 86th FIS; 47th FIS; 40th FIS; CT ANG
55-3426 323rd FIS; 86th FIS; 47th FIS; 40th FIS; CT ANG; Turkey
55-3427 5th FIS; 431st FIS; 4780th ADW; ND ANG; AZ ANG; MT ANG; ID ANG; drone
55-3428 31st FIS; 2nd FIS; 4th FIS; CT ANG; MASDC May 1970
55-3429 323rd FIS; 5th FIS; 4th FIS; CT ANG; Turkey
55-3430 323rd FIS; 325th FIS; 327th CAMS; 5th FIS; 4th FIS; CT ANG; MASDC May 1970
55-3431 323rd FIS; 86th FIS; 5th FIS; 431st FIS; 4780th ADW; ND ANG; MASDC Jan. 1970
55-3432 323rd FIS; 325th FIS; 327th CAMS; 5th FIS; 431st FIS; 4780th ADW; ND ANG; MASDC Nov. 1969
55-3433 31st FIS; 2nd FIS; 431st FIS; 4780th ADW; 4756th ADW; VT ANG; ND ANG; MT ANG; MASDC March 1970
55-3434 31st FIS; 2nd FIS; 431st FIS; 4780th ADW; WI ANG
55-3435 31st FIS; 5th FIS
55-3436 31st FIS; 5th FIS; 431st FIS; 4780th ADW; ND ANG; NMIMT
55-3437 323rd FIS; 325th FIS; 327th CAMS; 5th FIS; 431st FIS; 4780th ADW; 496th FIS; OR ANG; WA ANG; WI ANG
55-3438 323rd FTS; 86th FIS; 496th FIS; 2nd FIS; 431st FIS; 32nd FIS; 4780th ADW; WI ANG; crash/destroyed 3-5-68
55-3439 323rd FIS; destroyed by fire May 1957
55-3440 31st FIS; 5th FIS; 4th FIS; CT ANG; MASDC August 1970
55-3441 323rd FIS; crash/destroyed 3-19-57
55-3442 31st FIS; 5th FIS
55-3443 2nd FIS; 31st FIS; 4780th ADW; FL ANG; MT ANG; CA ANG; Turkey; crash/destroyed 8-21-68
55-3444 323rd FIS; 86th FIS; 2nd FIS; 496th FIS; 32nd FIS; 431st FIS; 498th FIS; 4780th ADW; 460th FIS; OR ANG; HI ANG; w/o 1967
55-3445 317th FIS; 2nd FIS; 496th FIS; 4780th ADW; 431st FIS; 525th FIS; 32nd FIS; ID ANG; crash/destroyed 5-20-67
55-3446 31st FIS; 5th FIS; 4th FIS; SD ANG; Turkey
55-3447 31st FIS; 5th FIS; 431st FIS; 525th FIS; 32nd FIS; ID ANG; drone
55-3448 31st FIS; 317th FIS; 2nd FIS; 496th FIS; 431st FIS; TX ANG; WA ANG; NMIMT
55-3449 31st FIS; 2nd FIS; 431st FIS; 4780th ADW; 4756th ADW; MT ANG; WI ANG; OR ANG; NY ANG; drone
55-3450 2nd FIS; 431st FIS; 4780th ADW; CT ANG; WI ANG; New England Air Museum
55-3451 31st FIS; 2nd FIS; 5th FIS; 431st FIS; 4780th ADW; AZ ANG; ND ANG
55-3452 323rd FIS; 31st FIS; 5th FIS; 4th FIS; CT ANG; Turkey; Burlington Airport, VT display
55-3453 2nd FIS; 5th FIS; 4th FIS; CA ANG; crash/destroyed 4-12-68
55-3454 86th FIS; 317th FIS; 2nd FIS; 431st FIS; 4780th ADW; 460th FIS; 460th FIS; 32nd FIS; OR ANG; HI ANG
55-3455 31st FIS; 2nd FIS; 4th FIS; MT ANG; CA ANG; WI ANG; Turkey
55-3456 2nd FIS; 431st FIS; 4780th ADW; 32nd FIS; WI ANG
55-3457 31st FIS; 2nd FIS
55-3458 317th FIS; 2nd FIS; 4th FIS; CA ANG; MASDC Nov. 1969
55-3459 4750th ADW/4750th TS; 2nd FIS
55-3460 31st FIS; 2nd FIS; 431st FIS; 4780th ADW; ND ANG; AZ ANG; NMIMT
55-3461 4750th ADW; 5th FIS; 2nd FIS; CA ANG; Turkey
55-3462 317th FIS; 2nd FIS; 4th FIS; CT ANG; Burlington Airport, VT display
55-3463 317th FIS; 2nd FIS; 4th FIS; MT ANG; CT ANG; MASDC 1970; NMIMT
55-3464 2nd FTS; 496th FIS; 431st FIS; 317th FIS; 317th FIS; WA ANG; MT ANG; MASDC March 1970
56-0957 31st FIS; 2nd FIS; 4th FIS
56-0958 317th FIS; 2nd FIS; 4th FIS
56-0959 2nd FIS; 5th FIS; 4th FIS; MT ANG; crash/destroyed 8-21-68
56-0960 31st FIS; 2nd FIS; 4th FIS; SD ANG
56-0961 31st FIS; 2nd FIS; 4th FIS; CA ANG; MASDC Jan. 1970
56-0962 2nd FIS; 4th FIS; CA ANG; MASDC Jan. 1970
56-0963 317th FIS; 5th FIS; 40th FIS; 509th FIS; crashed SVN 7-17-68
56-0964 317th FIS; 2nd FIS; 40th FIS; 509th FIS
56-0965 498th FIS; 5th FIS; 40th FIS; 509th FIS
56-0966 317th FIS; 5th FIS; crash/destroyed 6-8-59
56-0967 317th FTS; 2nd FIS; 5th FIS; 64th FIS; 509th FIS
56-0968 317th FIS; 5th FIS; 509th FIS

56-0969 2nd FIS; 5th FIS; 509th FIS
56-0970 2nd FIS; 5th FIS; 509th FIS; destroyed collision with F-4 Thailand
56-0971 317th FIS; 5th FIS; 40th FIS
56-0972 317th FIS; 2nd FIS; 509th FIS
56-0973 31st FTS; 86th FIS; 431st FIS; 32nd FIS; crash/destroyed 12-12-62
56-0974 31st FIS; 86th FIS; crash/destroyed 4-25-59
56-0975 317th FIS; 47th FIS 431st FIS; 32nd FIS; ID ANG; WA ANG; crash/destroyed 10-11-69
56-0976 317th FIS; crash/destroyed 2-25-57
56-0977 317th FIS; 86th FIS; 32nd FIS; ME ANG; FL ANG
56-0978 317th FIS; 86th FIS; HI ANG; AZ ANG; LA ANG; VT ANG; drone
56-0979 317th FIS; AZ ANG
56-0980 317th FIS; 47th FIS; 431st FIS; 32nd FIS; ME ANG; FL ANG; SC ANG; drone
56-0981 317th FIS; 86th FIS; 68th FIS; HI ANG; AZ ANG; Greece
56-0982 317th FIS; 318th FIS; 325th FIS; 32nd FIS; WA ANG; SC ANG; drone
56-0983 498th FIS; 48th FIS; 4756th ADG; 431st FTS; 497th FIS; 32nd FIS; ME ANG; FL ANG; drone
56-0984 498th FIS; Colorado museum
56-0985 317th FIS; 318th FIS; 325th FIS; 32nd FIS; ID ANG; WA ANG; SC ANG; McEntire ANGB display
56-0986 317th FIS; 318th FIS; 325th FIS; 32nd FIS; ME ANG; FL ANG; Clearwater Airport, FL display
56-0987 317th FIS; 86th FIS; 32nd FIS; ME ANG; FL ANG; SC ANG; LA ANG; drone
56-0988 31st FIS; 318th FIS; HI ANG; AZ ANG; Greece
56-0989 2nd FIS; 48th FIS; HI ANG; AZ ANG; CT ANG; NMIMT
56-0990 498th FIS; 86th FIS; 68th FIS; HI ANG; AZ ANG; SD ANG
56-0991 498th FIS; 86th FIS; 4780th ADW
56-0992 498th FIS; 32nd FIS; TX ANG; AZ ANG; HI ANG; NMIMT
56-0993 325th FIS; 323rd FIS; 318th FIS; 327th CAMS; 32nd FIS; ID ANG; WA ANG; SC ANG; drone
56-0994 317th FIS; 318th FIS; HI ANG; AZ ANG; CA ANG; MASDC May 1970
56-0995 Santa Maria Park, CA display
56-0996 2nd FIS; 86th FIS; 32nd FIS; WA ANG; ID ANG; CA ANG; MT ANG; drone
56-0997 2nd FIS; 32nd FIS
56-0998 325th FIS; 323rd FIS; 327th CAMS; 47th FIS; HI ANG; AZ ANG; CT ANG; SC ANG; NASA; MASDC June 1974
56-0999 2nd FIS
56-1000 498th FIS
56-1001 AFFTC/case XX wing tests; 48th FIS; TX ANG; Greece
56-1002 498th FIS; 86th FIS; 32nd FIS; ME ANG; FL ANG; crash/destroyed 3-23-71
56-1003 2nd FIS; 48th FIS; HI ANG; AZ ANG; CT ANG; ME ANG; drone
56-1004 2nd FIS; 48th FIS
56-1005 317th FIS
56-1006 2nd FIS; 48th FIS; 4780th ADW; 325th FIS; 431st FIS; 525th FIS; 32nd FIS; AZ ANG; ND ANG; SC ANG; drone
56-1007 2nd FIS; 48th FIS; TX ANG; Greece
56-1008 2nd FIS; 86th FIS; 525th FIS; 73rd AD; HI ANG; AZ ANG; CT ANG; MASDC June 1970
56-1009 2nd FIS; 48th FIS; HI ANG; AZ ANG; CA ANG; MASDC Jan. 1970
56-1010 498th FIS; crash/destroyed 1-24-58
56-1011 498th FIS; 48th FIS; AZ ANG; Greece
56-1012 325th FIS; 323rd FIS; 327th CAMS; 48th FIS; 431st FIS; crash/destroyed 12-12-63
56-1013 498th FIS; 86th FIS; 32nd FIS; ME ANG; FL ANG; drone
56-1014 318th FIS; 325th FIS; 32nd FIS; ME ANG; FL ANG; NY ANG; drone
56-1015 498th FIS; 48th FIS; TX ANG; crash/destroyed 1-12-61
56-1016 318th FIS; 325th CAMS; 48th FIS; AZ ANG; Greece
56-1017 2nd FIS; 48th FIS; HI ANG; AZ ANG; CA ANG; TX ANG; MASDC May 1970; Ellsworth AFB Museum
56-1018 325th FIS; 323rd FIS; 327th CAMS; 48th FIS; TX ANG; MASDC Jan. 1970
56-1019 318th FIS; 325th CAMS; 47th FIS; AZ ANG; TX ANG; MASDC Jan. 1970
56-1020 325th FIS; 323rd FIS; 327th CAMS; 47th FIS; crash/destroyed 9-10-58
56-1021 318th FIS; 325th CAMS; 47th FIS; 32nd FIS; crash/destroyed 9-25-61
56-1022 4750th ADW/4750th TS; crash/destroyed 1-7-58
56-1023 498th FIS; 4750th TS; 48th FIS; 431st FIS; 497th FIS; 32nd FIS; ME ANG; FL ANG; drone
56-1024 318th FIS; 325th CAMS; 48th FIS; TX ANG; Greece
56-1025 318th FIS; 325th CAMS; 47th FIS; 4780th ADW; CT ANG; Greece
56-1026 48th FIS; TX ANG
56-1027 2nd FIS; 318th FIS; 325th FIS; 32nd FIS; ME ANG; FL ANG
56-1028 498th FIS; 86th FIS; 32nd FIS; ME ANG; FL ANG; CT ANG; drone
56-1029 86th FIS; 32nd FIS; destroyed by fire 8-2-66
56-1030 2nd FIS; 48th FIS; HI ANG; AZ ANG; TX ANG; MASDC Jan. 1970
56-1031 2nd FIS; 48th FIS; MT ANG; HI ANG; CT ANG; Greece
56-1032 498th FIS; 86th FIS; 32nd FIS; 525th FIS; ME ANG; FL ANG; drone
56-1033 48th FIS; 431st FIS; HI ANG; AZ ANG; CA ANG; NMIMT
56-1034 48th FIS; HI ANG; AZ ANG; Greece
56-1035 2nd FIS; 48th FIS; HI ANG
56-1036 2nd FIS; 48th FIS; HI ANG; MT ANG; CA ANG; MASDC Jan. 1970
56-1037 2nd FIS; 48th FIS; TX ANG; MASDC Jan. 1970
56-1038 498th FIS; 48th FIS; TX ANG; CA ANG
56-1039 2nd FIS; 48th FIS; TX ANG; Greece
56-1040 48th FIS; TX ANG; Greece

56-1041 498th FIS; 318th FIS; HI ANG; CT ANG; NMIMT
56-1042 498th FIS; 86th FIS; 32nd FIS; 525th FIS; ND ANG; TX ANG; drone
56-1043 498th FIS; 86th FIS; 32nd FIS; 525th FIS; 526th FIS; ND ANG; TX ANG; drone
56-1044 498th FIS; 86th FIS; 32nd FIS; 525th FIS; ND ANG; TX ANG; drone
56-1045 318th FIS; 325th CAMS 82nd FIS; 497th FIS; 496th FIS; crash/destroyed 2-16-67
56-1046 325th FIS; 327th CAMS; 82nd FIS; 497th FIS; 431st FIS; 496th FIS; CA ANG; drone
56-1047 318th FIS; 325th CAMS; 87th FIS; 4780th ADW; 57th FIS; MASDC 7-24-7
56-1048 318th FIS; 431st FIS; 496th FIS; ID ANG; WA ANG; SC ANG; drone
56-1049 325th FIS; 327th CAMS; 87th FIS
56-1050 325th FIS; 327th CAMS; 82nd FIS; 497th FIS; 496th FIS; CA ANG drone
56-1051 5th FIS; destroyed by fire 2-19-58
56-1052 318th FIS; 325th CAMS; 87th FIS; 4780th ADW; 325th FTS; 431st FIS; CA ANG; Greece
56-1053 318th FIS; 325th CAMS; 82nd FIS; 497th FIS; 496th FIS; CA ANG; ID ANG; MASDC March 1975; Elmendorf AFB display
56-1054 5th FIS; 52nd CAMS; 87th FIS; ID ANG; drone
56-1055 317th FIS; 325th FIS; 327th CAMS; 82nd FIS; 4780th ADW; weapons test; ND ANG; ID ANG; MT ANG; NY ANG; drone
56-1056 5th FIS; 87th FIS; 4780th ADW; 325th FIS; TX ANG; Greece 56-1057 325th FIS; 327th CAMS; 82nd FIS; 497th FIS; 525th FIS; TD ANG; drone
56-1058 325th FIS; 327th CAMS; 82nd FIS; 327th FIS; 509th FIS; scrapped at Clark AB
56-1059 318th FIS; 325th CAMS; 87th FIS; 4780th ADW; CA ANG; Greece
56-1060 5th FIS; 52nd CAMS; 525th FIS; crash/destroyed 6-18-61
56-1061 325th FIS; 327th CAMS; 82nd FIS; 497th FIS; 496th FIS; CA ANG; ID ANG; drone
56-1062 325th FIS; 327th CAMS; 82nd FIS; 497th FIS; 496th FIS; 525th FIS; CA ANG; drone
56-1063 325th FIS; 327th CAMS; 82nd FIS; 497th FIS; 496th FIS; CA ANG; ID ANG; drone
56-1064 4756th ADG; 87th FIS; 4780th ADW; 317th FIS; 64th FIS; 509th FIS
56-1065 325th FIS; 327th CAMS 82nd FIS; 497th FIS; 496th FIS; CA ANG; crash/destroyed 3-10-7l
56-1066 325th FIS; 327th CAMS; 82nd FIS; crash/destroyed 9-20-58
56-1067 325th FIS; 327th CAMS; 82nd FIS; TX ANG
56-1068 318th FIS; 325th CAMS; crash/destroyed 4-9-58
56-1069 5th FIS; 87th FIS; 4780th ADW; LA ANG; FL ANG; drone
56-1070 4756th ADG; 87th FIS; crash/destroyed 1-5-60
56-1071 5th FIS; 86th FIS; 497th FIS; 525th FIS; crash/destroyed 12-14-65
56-1072 5th FIS; 82nd FIS; 497th FIS; 496th FIS; CA ANG; drone
56-1073 325th FIS; 327th CAMS; crash/destroyed 3-10-58
56-1074 48th FIS; 326th FIS; TX ANG; crash/destroyed 8-21-66
56-1075 5th FIS; 87th FIS
56-1076 5th FIS; 52nd CAMS; 525th FIS; ID ANG; crash/destroyed 3-20-72
56-1077 4756th ADG; 525th FIS; ID ANG; drone
56-1078 325th FIS; 327th CAMS; 82nd FIS; TX ANG; CA ANC; MASDC Jan. 1970
56-1079 318th FIS; 325th CAMS; 87th FIS; 4780th ADW; 325th FIS; TX ANG; Greece
56-1080 318th FIS; 325th CAMS; 82nd FIS; 431st FIS; 497th FIS; 496th FIS; CA ANG; crash/destroyed 7-27-71
56-1081 327th CAMS; 325th FIS; 82nd FIS; 497th FIS; 496th FIS; CA ANG; ID ANG; drone
56-1082 327th CAMS; 325th FIS; 82nd FIS; 497th FIS; 496th FIS; CA ANG; drone
56-1083 5th FTS; 52nd CAMS; 82nd FIS; 497th FIS; 496th FIS; 525th FIS; SC ANG; OR ANG; drone
56-1084 318th FTS; 325th CAMS; 87th FIS; 4780th ADW; crash/destroyed 9-24-63
56-1085 48th FIS; 86th FIS; crash/destroyed 1-2-59
56-1086 318th FIS; 325th CAMS; 82nd FIS; 497th FIS; 525th FIS; SC ANG; OR ANG; drone
56-1087 318th FIS; 325th CAMS; 87th FIS; 4780th ADW
56-1088 318th FIS; 325th CAMS; 82nd FIS
56-1089 318th FIS; crash/destroyed 4-24-57
56-1090 318th FIS; 82nd FIS; 325th CAMS; 497th FIS; 496th FIS; crash/destroyed 5-17-66
56-1091 5th FIS; 52nd CAMS; 87th FIS; 525th FIS; 4780th ADW; crash/destroyed 5-8-66
56-1092 318th FIS; 325th CAMS; 87th FIS; 59th FIS; 460th FIS; 525th FIS; OR ANG; HI ANG
56-1093 318th FIS; 325th CAMS; 86th FIS; 525th FIS; WI ANG; MT ANG; NY ANG; drone
56-1094 5th FIS; 52nd CAMS; 525th FIS; WI ANG
56-1095 318th FIS; 325th CAMS; 82nd FIS; 497th FIS; 496th FIS; 526th FIS; CA ANG; crash/destroyed 3-10-71
56-1096 318th FIS; 325th CAMS; 82nd FTS; 497th FIS; 496th FIS; CA ANG; drone
56-1097 318th FIS; 325th CAMS; 82nd FIS; 4780th ADW; 57th FIS; MASDC July 1970; NMIMT
56-1098 48th FIS; 86th FIS; 497th FIS; 496th FIS; CA ANG; ID ANG; drone
56-1099 5th FIS; 52nd CAMS; 86th FIS; 431st FIS; 496th FIS; CA ANG; PA ANG
56-1100 438th FIS; 87th FIS; 4780th ADW; MASDC 11-19-69
56-1101 48th FIS; 82nd FIS; 497th FIS; 496th FIS; CA ANG; NY ANG; drone
56-1102 318th FIS; 325th CAMS; 87th FIS; 4780th ADW; 325th FIS; 18th FIS; 57th FIS; MASDC Feb. 1971; NMIMT; Great Falls Airport, MT display

56-1103 482nd FIS; 86th FIS; crash/destroyed 1-6-60
56-1104 48th FIS; 71st FIS; shot down by another aircraft with GAR 9-14-59
56-1105 318th FIS; 325th CAMS; 82nd FIS; 497th FIS; 525th FIS; CA ANG; WI ANG; MT ANG; Great Falls Airport, MT display
56-1106 5th FIS; 52nd CAMS; 87th FIS; 4780th ADW; TX ANG; Greece; Ellsworth AFB display
56-1107 5th FIS; 52nd CAMS; 525th FIS; SC ANG; WA ANG; drone
56-1108 5th FIS; 52nd CAMS; 326th FIS; 4780th ADW; LA ANG; MASDC Oct. 1970
56-1109 482nd FIS; 48th FTS; 87th FIS; 3555th FTW; 4780th ADW; 57th FIS; MASDC Feb. 1971; Colorado museum
56-1110 438th FIS; 326th FIS 48th FIS; 4780th ADW; 317th FIS; 509th FIS; crash/destroyed SVN 1-19-66
56-1111 482nd FIS; 86th FIS; 32nd FIS; 525th FIS; SC ANG; WA ANG; drone
56-1112 438th FIS; 328th FG; 4780th ADW; MASDC 11-20-69
56-1113 438th FIS; 87th FIS; 48th FIS; 4780th ADW; 317th FIS; 509th FIS; scrapped
56-1114 5th FIS; 52nd CAMS; 86th FIS; 71st FIS; SD ANG; CA ANG; March Field Museum, CA
56-1115 5th FIS; 52nd CAMS; 87th FIS; 4780th ADW; LA ANG; Fairchild AFB display
56-1116 4756th ADW; 71st FIS; SD ANG; MT ANG; Helena College
56-1117 482nd FIS; 326th FIS; 4756th ADW; 4780th ADW; 57th FIS; 497th FIS; 496th FIS; CA ANG; MASDC Nov. 1969
56-1118 5th FIS; 52nd CAMS; 87th FIS; 4780th ADW; MASDC 11-25-69
56-1119 318th FIS; 87th FIS; TX ANG; MASDC Feb. 1970
56-1120 5th FIS; 52nd CAMS; 59th FIS; 525th FIS; destroyed by fire 11-8-65
56-1121 48th FIS; 82nd FIS; 497th FIS; 496th FIS; CA ANG; drone
56-1122 48th FIS; 86th FIS; 4756th ADG; 525th FIS; 32nd FIS; ME ANG; FL ANG; drone
56-1123 5th FIS; 52nd CAMS; 87th FIS; 509th FIS; TX ANG
56-1124 48th FIS; 326th FIS; SD ANG; CT ANG; MASDC May 1970
56-1125 5th FIS; 52nd CAMS; 87th FIS; 4780th ADW; SD ANG; Greece; Eifel Museum, Germany
56-1126 5th FTS; 52nd CAMS; 87th FIS; 4780th ADW; SD ANG; MASDC May 1970
56-1127 48th FIS; 82nd FIS; 482nd FIS; 4780th ADW; 317th FIS; 317th FIS; WA ANG; SC ANG; drone
56-1128 48th FIS; crash/destroyed 9-1-57
56-1129 482nd FIS; 326th FIS; TX ANG; MASDC Jan. 1970
56-1130 5th FIS; 52nd CAMS; 525th FIS; 32nd FIS; ME ANG; FL ANG; drone
56-1131 48th FIS; 86th FIS; 496th FIS; 525th FIS; VT ANG; TX ANG; crash/destroyed 9-24-70
56-1132 482nd FIS; 71st FIS; 1st CAMS; 4780th ADW; 57th FIS; MASDC 1-27-71
56-1133 48th FIS; 325th CAMS; 1st CAMS; SD ANG; MASDC June 1970
56-1134 482nd FIS; 86th FIS; 68th FIS; 71st FIS; SD ANG; LA ANG; AZ ANG; Tucson Airport, AZ display
56-1135 5th FTS; 52nd CAMS; 87th FIS; 4780th ADW; WA ANG; SC ANG; drone
56-1136 48th FIS; 82nd FIS; 497th FIS; 496th FIS; CA ANG; PA ANG; drone
56-1137 438th FIS; crash/destroyed 6-6-57
56-1138 82nd FIS; 326th FIS; 4780th ADW; CA ANG; NMIMT
56-1139 438th FIS; w/o accident 11-1-57
56-1140 438th FIS; 95th FIS; 509th FIS; HI ANG; McClellan Museum, CA
56-1141 525th FIS; 326th FIS; ,crash/destroyed 6-10-59
56-1142 326th FIS; 4780th ADW; crash/destroyed 7-19-62
56-1143 48th FIS; 438th FIS; 95th FIS
56-1144 438th FIS; 95th FIS; 509th FIS
56-1145 326th FIS; 4780th ADW; MASDC 12-2-69
56-1146 48th FIS; 438th FIS; 95th FIS; 16th FIS; 509th FIS; w/o Clark AB 1-31-67
56-1147 48th FIS; 438th FIS; 16th FIS; 95th FIS; HI ANG
56-1148 326th FIS; crash/destroyed 1-25-58
56-1149 326th FIS; 4780th ADW; LA ANG; CT ANG
56-1150 482nd FIS; 95th FIS; 509th FIS
56-1151 48th FIS; 438th FTS; 95th FIS; Lackland AFB, TX display
56-1152 326th FIS; 482nd FIS; 95th FIS; HI ANG
56-1153 326th FIS; 4780th ADW; MASDC 11-14-69
56-1154 438th FIS; 95th FIS; 16th FIS
56-1155 326th FIS; 438th FIS; SD ANG; crash/destroyed 8-20-64
56-1156 482nd FIS; 95th FIS; 509th FIS
56-1157 48th FIS; 95th FIS; 438th FIS; 509th FIS
56-1158 482nd FIS; 95th FIS; 509th FIS
56-1159 48th FIS; 4780th ADW; LA ANG; CA ANG; MASDC Jan. 1970
56-1160 48th FIS; 438th FIS; crash/destroyed 1-16-58
56-1161 482nd FIS; 95th FIS; 509th FIS; destroyed ground attack SVN 7-1-65
56-1I62 482nd FIS; 95th FIS; 64th FIS; HI ANG; 509th FIS; crash/destroyed SVN 12-9-66
56-1163 27th FIS; 526th FIS; 32nd FIS; 525th FIS; ME ANG; FL ANC; drone
56-1164 82nd FIS; 326th FIS; crash/destroyed 5-29-59
56-1165 48th FTS; 438h FIS; 509th FIS; destroyed ground attack SVN 5-12-61
56-1166 438th FIS; 95th FIS; 509th FIS; shot down by MiG SEA
56-1167 438th FIS; 95th FIS; SD ANG; HI ANG
56-1168 326th FIS; 4780th ADW; MASDC 11-17-69
56-1169 48th FIS; 438th FIS; 95th FIS
56-1170 82nd FIS; 27th FIS; 526th FIS; ME ANG; FL ANG; drone
56-1171 27th FIS; crash/destroyed 6-28-59
56-1172 438th FIS; 95th FIS; 1st CAMS; 71st FIS; 317th FIS; crash/destroyed 11-20-68

56-1173 326th FIS; crash/destroyed 12-16-57
56-1174 326th FIS; 71st FIS; 1st CAMS; SD ANG
56-1175 27th FIS; 526th FIS; ME ANG; FL ANG; drone
56-1176 438th FIS; 95th FIS; 71st FIS; 1st CAMS; 525th FIS
56-1177 438th FIS; 95th FIS; 71st FIS; 1st CAMS; 16th FIS; HI ANG
56-1178 27th FIS; 526th FIS; 496th FIS; 525th FIS; crash/destroyed 10-18-64
56-1179 326th FIS; 1st CAMS; 71st FIS; crash/destroyed 7-1-59
56-1180 326th FIS; 1st CAMS; 71st FIS; 317th FIS; crash/destroyed 1-14-64
56-1181 438th FIS; 95th FIS; 71st FIS; 1st CAMS; 317th FIS; WA ANG; SC ANG; drone
56-1182 326th FIS; 1st CAMS; 71st FIS; 509th FIS; destroyed ground attack SVN 7-1-65
56-1183 27th FIS; 71st FIS; 1st CAMS
56-1184 27th FIS; 37th FIS; 526th FIS; 509th FIS; SC ANG; drone 56-1185 326th FIS; 4780th ADW; MASDC 6-12-70
56-1186 482nd FIS; 95th FIS; 16th; 509th FIS; crash/destroyed Thailand 1-8-69
56-1187 27th FIS; 71st FIS; 1st CAMS; SD ANG; CT ANG; NMIMT
56-1188 27th FIS; 526th FIS; 525th FIS; VT ANG; TX ANG; NMIMT; drone
56-1189 326th FIS; 71st FIS; 1st CAMS; 64th FIS; 509th FIS; crash/destroyed SVN 11-24-64
56-1190 82nd FIS; 326th FIS; 497th FIS; 4780th ADW; LA ANG; CA ANG; MASDC Jan. 1970
56-1191 326th FIS; 71st FIS; 1st CAMS; 57th FIS; 76th FIS; 4780th ADW; OR ANG; HI ANG
56-1192 482nd FIS; 95th FIS; 509th FIS; HI ANG
56-1193 438th FIS; 95th FIS; 1st CAMS; 64th FIS; 482nd FIS; 4780th ADW; WA ANG; SC ANG; drone
56-1194 326th FIS; 71st FIS; 1st CAMS; crash/destroyed 5-4-60
56-1195 326th FIS; 1st CAMS; 71st FIS; 317th FIS; crash/destroyed 2-3-61
56-1196 5th FIS; 1st CAMS; 71st FIS; 317th FIS; 11th FIS; WA ANG; SC ANG; drone
56-1197 326th FIS; 4780th ADW; crash/destroyed early 1966
56-1198 82nd FIS; 326th FIS; 4780th ADW; TX ANG; CA ANG; MASDC Jan. 1970
56-1199 482nd FIS
56-1200 482nd FIS; 95th FIS; 509th FIS
56-1201 482nd FIS; 95th FIS; SD ANG; MASDC May 1970
56-1202 27th FIS; 37th FIS; 526th FIS; MT ANG; NY ANG; PA ANG; drone
56-1203 82nd FIS; 326th FIS; 4780th ADW; MT ANG; LA ANG; CA ANG; NMIMT
56-1204 326th FIS; 1st CAMS; 525th FIS; 71st FIS; crash/destroyed 4-29-59
56-1205 326th FIS; 71st FIS; 317th FIS; 482nd FIS; 4780th ADW; WI ANG; drone
56-1206 482nd FIS; 95th FIS; 317th FIS; 526th FIS; 525th FIS; VT ANG; TX ANG; drone
56-1207 27th FIS; 71st FIS; 1st CAMS
56-1208 82nd FIS; 27th FIS; 526th FIS; MT ANG; NY ANG; drone
56-1209 82nd FIS; 326th FIS; 4780th ADW; TX ANG
56-1210 27th FIS; 525th FIS; 526th FIS; 431st FIS; MT ANG; SC ANG; NY ANG; drone
56-1211 27th FIS; 526th FIS; 525th FIS; 32nd; MT ANG; SC ANG; NY ANG; drone
56-1212 82nd FIS; 326th FIS; 4780th ADW; MASDC 11-26-69
56-1213 27th FIS; 71st FIS; 1st CAMS; 497th FIS; 317th FIS; WA ANG; SC ANG; drone
56-1214 27th FIS; 37th FIS; 526th FIS; 497th FIS; 525th FIS; destroyed landing accident 5-13-66
56-1215 31st FIS; 326th FIS; 71st FIS; 1st CAMS; 317th FIS; VT ANG; MT ANG; NY ANG; WA ANG; drone
56-1216 27th FIS; 526th FIS; 525th FIS; VT ANG; MT ANG; MN ANG
56-1217 27th FIS; 526th FIS; 525th FIS; MN ANG
56-1218 482nd FIS; 95th FIS; 526th FIS; 509th FIS; HI ANG; TX ANG; crash/destroyed 6-15-73
56-1219 27th FIS; 526th FIS; 496th FIS; CT ANG; Minneapolis, MN museum
56-1220 27th FIS; 438th FIS; 95th FIS; 71st FIS; 1st CAMS; 525th FIS; 317th FIS; 509th FIS
56-1221 482nd FIS; 95th FIS; 526th FIS; CT ANG; New England Air Museum
56-1222
56-1223 27th FIS; 526th FIS; CT ANG; CA ANG; NY ANG; drone
56-1224 27th FIS; crash/destroyed 11-25-57
56-1225 27th FIS; 71st FIS; 1st CAMS; 317th FIS; crash/destroyed 9-10-62
56-1226 82nd FIS; 326th FIS; 317th FIS; 509th FIS; scrapped
56-1227 27th FIS; 526th FIS; CT ANG; ID ANG; drone
56-1228 82nd FIS; 326th FIS; 431st FIS; 4780th ADW; LA ANG; MASDC Nov. 1969
56-1229 27th FIS; 71st FIS; 1st CAMS; 326th FIS; TX ANG; NMIMT
56-1230 82nd FIS; 27th FIS; SD ANG; MASDC May 1970
56-1231 27th FIS; 438th FIS; 95th FIS; 509th FIS
56-1232 27th FIS; 326th FIS; 4780th ADW; LA ANG; Greece
56-1233 82nd FIS; 326th FIS; 431st FIS; ATC; 4780th ADW; LA ANG; NY ANG; Greece
56-1234 332nd FIS; 482nd FIS 27th FIS; 526th FIS; 525th FIS; crash/destroyed 3-5-68
56-1235 64th FIS; 325th CAMS; 82nd FIS; 27th FIS; 526th FIS; crash/destroyed 3-5-63
56-1236 332nd FIS; 482nd FIS; 27th FIS; 526th FIS; 32nd FIS; 525th FIS; MT ANG; NY ANG; ID ANG; CA ANG; drone
56-1237 332nd FIS; 482nd FIS; 27th FIS; 526th FIS; crash/destroyed 7-14-69
56-1238 332nd FIS; 482nd FIS; 27th FIS; 526th FIS; CT ANG; MASDC June 1971
56-1239 82nd FIS; 27th FIS; 526th FIS; crash/destroyed 1-6-61
56-1240 64th FIS; 325th CAMS; 82nd FIS; 27th FIS; 526th FIS; CT .ANG; MASDC 1971

56-1241 332nd FIS; 482nd FIS; 27th FIS; 526th FIS; CT ANG; CA ANG; NY ANG; drone
56-1242 317th FIS; 525th FIS; 526th FIS; CT ANG; MASDC June 1971
56-1243 317th FIS; 525th FIS; crash/destroyed 11-26-60
56-1244 317th FIS; 525th FIS; 526th FIS; 32nd FIS; ME ANG; ID ANG; CT ANG; FL ANG
56-1245 317th FIS; 525th FIS; 526th FIS; 32nd FIS; WI ANG; SC ANG; TX ANG; crash/destroyed 3-24-70
56-1246 317th FIS; crash/destroyed 5-1-58
56-1247 31st FIS; 4780th ADW; 460th FIS; 82nd FIS; 525th FIS; 526th FIS; 32nd FIS; OR ANG; HI ANG; Travis AFB display
56-1248 525th FIS; 317th FIS; WI ANG; drone
56-1249 317th FIS; 525th FIS; 526th FIS; CT ANG; PA ANG; CA ANG; drone
56-1250 323rd FIS; 4780th ADW; 57th FIS; drone
56-1251 317th FIS; crash/destroyed 10-16-63
56-1252 TX ANG; Ellington AFB display
56-1253 31st FIS; 317th FIS; 525th FIS; 526th FIS; CT ANG; ID ANG; drone
56-1254 317th FIS; WI ANG; drone
56-1255 317th FIS; 525th FIS; 526th FIS; CT ANG; ID ANG; drone
56-1256 31st FIS; 525th FIS; CT ANG; ID ANG; drone
56-1257 317th FIS; WI ANG; NY ANG; drone
56-1258 317th FIS; 525th FIS; 526th FIS; CT ANG; MASDC May 1971
56-1259 525th FIS; 317th FIS; TX ANG; drone
56-1260 317th FIS; 525th FIS; 526th FIS; CT ANG; TX ANG
56-1261 317th FIS; 525th FIS; 526th FIS; 496th FIS; SC ANG; drone
56-1262 317th FIS; TX ANG; drone
56-1263 317th FIS; 525th FIS; 526th FIS; CT ANG; ID ANG; drone
56-1264 31st FIS; 317th FIS; 525th FIS; 526th FIS; CT ANG; became GF-102A; Bradley Airport, CT display
56-1265 137th FIS; 525th FIS; 526th FIS; CT ANG; MASDC June 1971
56-1266 317th FIS; 525th FIS; 526th FIS; CT ANG; Stephenville, Newfoundland display
56-1267 317th FIS; WI ANG; drone
56-1268 31st FIS; 317th FIS; WI ANG; MASDC July 1971; Kelly AFB, TX display
56-1269 31st FIS; 317th FIS; WI ANG; drone
56-1270 317th FIS; 331st FIS; WI ANG; 509th FIS; drone
56-1271 317th FIS; crash/destroyed 9-1-57
56-1272 31st FIS; 317th FIS; WA ANG; SC ANG; drone
56-1273 31st FIS; 317th FIS; WI ANG; Volk Field, WI display
56-1274 31st FIS; 317th FIS; WI ANG; NY ANG; drone; Elmendorf AFB, AK display
56-1275 31st FIS; 317th FIS; crash/destroyed 5-29-62
56-1276 31st FIS; 317th FIS; crash/destroyed 1-2-62
56-1277 526th FIS; 31st FIS; 317th FIS; WI ANG; NY ANG; drone
56-1278 31st FIS; 317th FIS; WI ANG; drone
56-1279 31st FIS; 317th FIS; WI ANG
56-1280 31st FIS; 317th FIS; crash/destroyed 11-16-59
56-1281 31st FIS; 317th FIS; crash/destroyed 12-27-63
56-1282 31st FIS; 317th FIS; Wasilla, AK museum
56-1283 31st FIS; 317th FIS; crash/destroyed 6-13-69
56-1284 31st FIS; 317th FIS; WI ANG; drone
56-1285 323rd FIS; SD ANG; MASDC June 1970
56-1286 31st FIS; 317th FIS; crash/destroyed 10-27-62
56-1287 31st FIS; 317th FIS; WI ANG; drone
56-1288 323rd FIS; LA ANG; MASDC Mar. 1971
56-1289 31st FIS; 317th FIS; TX ANG; drone
56-1290 31st FIS; 317th FIS; WI ANG; drone
56-1291 431st FIS; 31st FIS; 317th FIS; WI ANG; drone
56-1292 31st FIS; 317th FIS; TX ANG; drone
56-1293 323rd FIS; SD ANG; drone
56-1294 31st FIS; 317th FIS; WI ANG; drone
56-1295 31st FIS; 317th FIS; TX ANG; drone
56-1296 323rd FIS; 137th FIS; LA ANG; MASDC Nov. 1970
56-1297 323rd FIS; 317th FIS; LA ANG; crash/destroyed 6-4-63
56-1298 323rd FIS; LA ANG; MASDC Oct. 1970
56-1299 323rd FIS; LA ANG; MASDC Jan. 1971
56-1300 31st FIS; 323rd FIS; 317th FIS; SD ANG; SC ANG; drone
56-1301 323rd FIS; 117th FIS; crash/destroyed 3-8-63
56-1302 323rd FIS; SD ANG; MASDC June 1970
56-1303 323rd FIS; SD ANG; MASDC May 1970
56-1304 323rd FIS; LA ANG; MASDC Jan. 1971
56-1305 323rd FIS; LA ANG
56-1306 323rd FIS; LA ANG; MASDC Dec. 1970; NMIMT
56-1307 323rd FIS; LA ANG; TX ANG; crash/destroyed mid-air with 51-1311 on 9-14-61
56-1308 323rd FIS; 329th FIS; LA ANG; MASDC Nov. 1970
56-1309 323rd FIS; PA ANG; TX ANG; crash/destroyed 2-4-61
56-1310 323rd FIS; LA ANG; NMIMT
56-1311 323rd FIS; TX ANG; crash/destroyed mid-air with 56-1307 on 9-14-61
56-1312 323rd FIS; LA ANG
56-1313 323rd FIS; crash/destroyed 4-23-59
56-1314 323rd FIS; 4780th ADW; 57th FIS; LA ANG; TX ANG; ID ANG; drone
56-1315 323rd FIS; LA ANG; crash/destroyed 9-20-66

56-1316 323rd FIS; 431st FIS; SD ANG; MASDC May 1970
56-1317 prototype for Case XX wing; 318th FIS; 325th CAMS; 64th FIS; AFFTC; 4780th ADW; 4756th ADW; FL ANG; CA ANG
56-1318 4756th ADW; 482nd FIS; 95th FIS; 326th FIS; 57th FIS; 64th FIS; w/o Clark AB 1-31-67
56-1319 82nd FIS; 4780th ADW; 4756th ADW; 57th FIS; drone
56-1320 64th FIS; 325th CAMS; 82nd FIS; 326th FIS; 4756th ADW; ADWC; PA ANG
56-1321 4756th ADW; 27th FIS; 4780th ADW; 57th FIS; crash/destroyed 1-22-73
56-1322 82nd FIS; 4756th ADW; 4780th ADW; ADWC; MASDC 11-12-69
56-1323 332nd FIS; 601st CAMS; 331st FIS; 328th FG; 326th FIS; WI ANG; MN ANG; CA ANG; drone
56-1324 332nd FIS; 460th FIS; 4683rd ADW; crash/destroyed 9-21-63
56-1325 332nd FIS; 59th FIS; MN ANG; CA ANG; drone
56-1326 332nd FIS; 331st FIS; 326th FIS; CA ANG; 328th FW; crash/destroyed 3-29-65
56-1327 332nd FIS; 4683rd ADW; 326th FIS; 4756th ADW; crash/destroyed 7-14-6
56-1328 332nd FIS; 82nd FIS; 326th FIS; 57th FG; 64th FIS; 509th FIS
56-1329 601st CAMS; 460th FIS; 326th FIS; 326th FIS; 4756th ADW; ADWC; CA ANG; PA ANG; drone
56-1330 332nd FIS; 4780th ADW; PA ANG; ND ANG; drone
56-1331 332nd FIS; 4780th ADW; MN ANG; PA ANG; drone
56-1332 64th FIS; 318th FIS; 325th CAMS; AFSWC; 57th FG; 326th FIS; 4780th ADW
56-1333 64th FIS; 318th FIS; 325th CAMS; 57th FG; 509th FIS
56-1334 332nd FIS; 4683rd ADW; 4780th ADW; WI ANG; MN ANG; MASDC April 1971
56-1335 332nd FIS; 325th CAMS; 326th FIS; 57th FG; 64th FIS; 82nd FIS; 509th FIS; scrapped Naha AB 1971
56-1336 332nd FIS; 59th FIS; MN ANG; CA ANG; drone
56-1337 332nd FIS; 326th FIS; 4756th ADW; ADWC; CA ANG; OR ANG; drone
56-1338 64th FIS; 325th CAMS; 57th FG; scrapped Naha AB 1971
56-1339 64th FIS; SC ANG; TX ANG; crash/destroyed 1-23-59
56-1340 64th FIS; 325th CAMS; 4756th ADW; ADWC; PA ANG; CT ANG; TX ANG; drone
56-1341 4756th ADW; 57th FIS; NY ANG; drone
56-1342 332nd FIS; 82nd FIS; 326th FIS; 57th FG; 64th FIS; scrapped Naha AB 1971
56-1343 325th CAMS; 57th FG; 325th FIS; 326th FIS; 64th FIS; 4780th ADW; WI ANG; drone
56-1344 64th FIS; 325th CAMS; crash/destroyed 2-8-64
56-1345 64th FIS 325th CAMS; 76th FIS; 57th FG; 59th FIS; FL ANG; WI ANG; MN ANG; drone
56-1346 64th FIS; 325th CAMS; 326th FIS; 4780th ADW; PA ANG; FL ANG; drone
56-1347 64th FIS; 325th CAMS; 326th FIS; 4780th ADW; SAALC; PA ANG; drone
56-1348 61st FIS; 327th CAMS; crash/destroyed 5-5-58
56-1349 64th FIS; 325th CAMS; 327th FG; 326th FIS; 4780th ADW; MT ANG NY ANG; CA ANG; drone
56-1350 64th FIS; 325th CAMS; 4756th ADG; 57th FIS; 318th FIS; drone
56-1351 61st FIS; 327th CAMS; PA ANG; CA ANG; drone
56-1352 64th FIS; 325th CAMS; 76th FIS
56-1353 4756th ADW; 76th FIS; 482nd FIS; crash/destroyed 2-23-62
56-1354 61st FIS; 327th CAMS; FL ANG; MASDC Nov. 1969
56-1355 4756th ADW; 57th FIS; w/o 12-19-69
56-1356 325th FIS; 61st FIS; 327th CAMS; 4780th ADW; 57th FIS; FL ANG; SC ANG; drone
56-1357 61st FIS; 325th FIS; 327th CAMS; WI ANG; FL ANG; crash/destroyed 12-2-69
56-1358 460th FIS; TN ANG; FL ANG; crash/destroyed 10-11-65
56-1359 327th FIS; 4756th ADW; 76th FIS; 460th FIS; crash/destroyed 10-22-64
56-1360 327th FIS; 4756th ADW; 76th FIS; 460th FIS; 82nd FIS; OR ANG; CA ANG; PA ANG; drone
56-1361 327th FIS; 4756th ADW; 76th FIS; 460th FIS; 64th FIS; 326th FIS; 4780th ADW; TX ANG
56-1362 327th FIS; 4756th ADW; 76th FIS; 64th FIS
56-1363 327th FIS; 4756th ADW; 76th FIS; 460th FIS; OR ANG; CA ANG; drone
56-1364 327th FIS; 73rd AD; 76th FIS; crash/destroyed 8-17-62
56-1365 327th FIS; 4756th ADW; 76th FIS; 326th FIS; ND ANG; PA ANG; Syracuse Park, NY display
56-1366 327th FIS; 4756th ADW; 95th FIS; 76th FIS; 460th FIS; OR ANG; CA ANG; PA ANG; drone
56-1367 327th FIS; 4756th ADW; 76th FIS; 64th FIS; 59th FIS; 57th FIS; ADWC; FL ANG; drone
56-1368 327th FIS; 4756th ADW; 76th FIS; 460th FIS; OR ANG; Portland, OR museum
56-1369 332nd FIS; 4683rd ADW; 4780th ADW; MT ANG; OR ANG
56-1370 61st FIS; 327th FIS; PA ANG; crash/destroyed 3-16-63
56-1371 332nd FIS; 64th FIS; HI ANG; crash/destroyed 7-12-62
56-1372 332nd FIS; 326th FIS; 64th FIS
56-1373 332nd FIS; 326th FIS; 4756th ADW; ADWC; FL ANG; drone
56-1374 332nd FIS; 82nd FIS; 326th FIS; 4756th ADW; ADWC
56-1375 332nd FIS; 325th CAMS; 326th FIS; 4756th ADW; crash/destroyed mid-air with 56-1383 on 12-22-66
56-1376 332nd FIS; 4683rd ADW; 325th CAMS; 4756th ADW; ADWC; LA ANG; FL ANG; SC ANG; drone
56-1377 332nd FIS; 460th FIS; 326th FIS; crash/destroyed 1-25-66
56-1378 4756th ADW; 57th FIS; w/o 6-11-71; NAS Keflavik, Iceland display
56-1379 64th FIS; 325th CAMS; 509th FIS; crash/destroyed 4-21-62

56-1380 4756th ADW; 332nd FIS; 326th FIS; 4683rd ADW; 59th FIS; MN ANG; VT ANG; PA ANG; drone
56-1381 332nd FIS; 4683rd ADW; crash/destroyed 7-14-61
56-1382 332nd FIS; 331st FIS; 326th FIS; ND ANG; MT ANG; CA ANG; NY ANG; crash/destroyed 12-1-73
56-1383 326th FIS; 460th FIS; 332nd FIS; 4756th ADW; crash/destroyed mid-air with 56-1375 on 12-22-66
56-1384 64th FIS; 325th CAMS; 4756th ADW; 496th FIS; 4780th ADW; MT ANG; NY ANG; ID ANG; drone
56-1385 327th FIS; 64th FIS; 325th CAMS
56-1386 61st FIS; 327th CAMS; FL ANG; crash/destroyed 7-19-65
56-1387 325th CAMS; 64th FIS; crash/destroyed 11-10-62
56-1388 327th FIS; PA ANG; MASDC Nov. 1969
56-1389 325th CAMS; 64th FIS; shot down SVN 12-13-66
56-1390 61st FIS; 327th CAMS; FL ANG; MASDC Nov. 1969
56-1391 64th FIS; 325th CAMS; 326th FIS; 57th FIS; 4780th ADW; CA ANG; ID ANG; drone
56-1392 4756th ADW; FL ANG; MASDC Nov. 1969
56-1393 4756th ADW; 4780th ADW; FL ANG; CT ANG; Pima museum, AZ
56-1394 4756th ADW; 57th FIS; MASDC 5-10-71
56-1395 64th FIS; 325th CAMS; crash/destroyed 12-26-57
56-1396 318th FIS; 4756th ADW; ADWC; 73rd AD; 57th FIS; CA ANG; crash/destroyed 3-25-68
56-1397 64th FIS; 325th CAMS; 76th FIS; 460th FIS; 82nd FIS; scrapped Naha AB
56-1398 64th FIS; 325th CAMS; 326th FIS; 4780th ADW; PA ANG; drone
56-1399 61st FIS; 59th FIS; 325th CAMS; 327th CAMS; FL ANG; crash/destroyed 8-14-64
56-1400 61st FIS; 327th CAMS; SAALC; PA ANG; CT ANG; CA ANG; drone
56-1401 61st FIS; 327th CAMS; 325th FIS; 57th FIS; 4780th ADW; TX ANG; PA ANG; drone
56-1402 61st FIS; 327th CAMS; FL ANG; crash/destroyed 3-20-69
56-1403 4756th ADW; 57th FIS; crash/destroyed 9-15-66
56-1404 61st FIS; 327th CAMS; crash/destroyed 5-10-60
56-1405 61st FIS; 327th CAMS; PA ANG; FL ANG; MASDC Nov. 1969; SAC museum, NE
56-1406 61st FIS; 327th CAMS; FL ANG
56-1407 61st FIS; 327th CAMS; PA ANG
56-1408
56-1409 61st FIS; 327th CAMS; SC ANG; PA ANG; FL ANG; MN ANG; MASDC Feb. 1970
56-1410 61st FIS; 327th CAMS; PA ANG; CA ANG; MASDC April 1970
56-1411 57th FIS; 4780th ADW; w/o Jan. 1970; became instructional airframe at Perrin AFB
56-1412 61st FIS; 327th CAMS; FL ANG; MASDC Nov. 1969; NMIMT
56-1413 61st FTS; 59th FIS; 327th CAMS; PA ANG; Chino, CA museum
56-1414 61st FIS; 327th CAMS; FL ANG; crash/destroyed 5-18-67
56-1415 61st FIS; 327th CAMS; 327th CAMS; PA ANG; Pittsburgh Airport display
56-1416 4756th ADW; 4780th ADW; 57th FIS; USAF museum
56-1417 4756th ADW; 57th FIS; AFSC; destroyed July 1974 in drone explosive charge test
56-1418 61st FIS; 327th CAMS; 4780th ADW; 57th FIS; FL ANG; CA ANG; drone
56-1419 4756th ADW; 57th FIS; drone
56-1420 64th FIS; 325th CAMS; 327th FIS; 318th FIS
56-1421 61st FIS; 327th CAMS; FL ANG; MASDC Nov. 1969
56-1422 73rd AD; 4750th TS; 76th FIS; 64th FIS; crash/destroyed 12-2-64
56-1423 61st FIS; 327th FIS; 327th CAMS; PA ANG; crash/destroyed 10-29-62
56-1424 18th FIS; 327th FIS; 412th CAMS; 87th FIS; PA ANG; MASDC Nov. 1969
56-1425 64th FIS; 325th CAMS; crash/destroyed 10-15-63
56-1426 4756th ADW; 4750th TS; 76th FIS; 64th FIS; 4780th ADW; 57th FIS; 326th FIS; ID ANG; CA ANG; drone
56-1427 18th FIS; 332nd FIS; 325th FIS; 32nd FIS; WI ANG; VT ANG; SC ANG; drone
56-1428 18th FIS; PA ANG; NMIMT
56-1429 61st FIS; 327th CAMS; FL ANG; CA ANG; MASDC April 1970
56-1430 37th FIS; 18th FIS; 11th FTS; 331st FIS; 326th FIS; 4756th ADW; CA ANG; drone
56-1431 18th FIS; ND ANG; TX ANG
56-1432 18th FIS; 438th FIS; 323rd FIS; 59th FIS; MN ANG; MASDC May 1971; Camp Robinson, AR display
56-1433 18th FIS; VT ANG; TX ANG; drone
56-1434 438th FIS; 59th FIS; MN ANG; PA ANG; VT ANG; CA ANG; drone
56-1435 37th FIS; 18th FIS; crash/destroyed 3-2-59
56-1436 438th FIS; 323rd FG; 59th FIS; 327th FG; 4780th ADW; 509th FIS
56-1437 18th FIS; 87th FIS; 323rd FIS; 59th FIS; CA ANG
56-1438 18th FIS; CT ANG; TX ANG; MT ANG; MASDC May 1970
56-1439 438th FIS; 59th FIS; 327th FG; 4780th ADW; 325th FIS; drone
56-1440 18th FIS; 482nd FIS; 76th FIS; 82nd FIS; scrapped at Naha AB 1971
56-1441 18th FIS; TX ANG; crash/destroyed 5-31-63
56-1442
56-1443 438th FIS; 59th FIS; ADWC; NY ANG; VT ANG; MN ANG; drone
56-1444 18th FIS; 482nd FIS; 76th FIS; 326th FIS; 64th FIS; 509th FIS
56-1445 18th FIS
56-1446 438th FIS; 323rd FIS; 59th FIS; 325th FIS; 4780th ADW; WI ANG; drone

56-1447 18th FIS; 57th FIS; 59th FIS; 4780th ADW; TX ANG; CA ANG; FL ANG; ID ANG; drone
56-1448 18th FIS; WA ANG; TX ANG; CA ANG; MASDC Jan. 1970
56-1449 18th FIS; 59th FIS; 327th FG; 4780th ADW; SC ANG; drone
56-1450 18th FIS; 326th FIS; 64th FIS
56-1451 18th FIS; 326th FIS; 64th FIS; 82nd FIS; scrapped at Naha AB 1971
56-1452 438th FIS; 59th FIS; crash/destroyed 8-15-60
56-1453 438th FIS; 323rd FIS; 59th FIS; MN ANG; CA ANG; drone
56-1454 18th FIS; 482nd FIS; 76th FIS; crash/destroyed 5-3-62
56-1455 18th FIS; 64th FIS; 326th FIS; 4756th ADW; 57th FIS; MASDC 5-13-71
56-1456 18th FIS; 64th FIS; 326th FIS; 4756th ADG; ADWC; PA ANG; drone
56-1457 438th FIS; 323rd FIS; 325th FIS; 59th FIS; ND ANG; PA ANG; drone
56-1458 18th FIS; 482nd FIS; 76th FIS; 460th FIS; crash/destroyed 11-20-64
56-1459 18th FIS; TX ANG
56-1460 438th FIS; 59th FIS; MN ANG; CA ANG; NY ANG; MT ANG; CT ANG; VT ANG; drone
56-1461 11th FIS; 331st FIS; 460th FIS; OR ANG; PA ANG
56-1462 18th FTS; 318th FIS; LA ANG; FL ANG; TX ANG; MASDC Jan. 1971
56-1463 438th FIS; 59th FIS; 325th FIS; 64th FIS; 82nd FIS; scrapped at Naha AB 1971
56-1464 438th FIS; 59th FIS; 460th FIS; 327th FG; 4780th ADW; TX ANG; drone
56-1465 327th FIS; 18th FIS; PA ANG; FL ANG; CA ANG; NMIMT
56-1466 37th FIS; 438th FIS; 498th FIS; 331st FIS; 460th FIS; OR ANG; CA ANG; drone
56-1467 11th FIS; 331st FIS; 4780th ADW; 327th FG; 64th FIS; 509th FIS
56-1468 438th FIS; crash/destroyed 6-5-59
56-1469 4756th ADW; 59th FIS; 325th FIS; 326th FIS; TX ANG; 4780th ADW; scrapped Naha AB 1971
56-1470 11th FIS; 331st FIS; 460th FIS; 82nd FIS; scrapped Naha AB 1971
56-1471 11th FIS; 331st FIS; 4756th ADW; ADWC; FL ANG; drone
56-1472 18th FIS; 59th FIS; 325th FIS; 4780th ADW; WI ANG; FL ANG; PA ANG; drone
56-1473 37th FIS; 438th FIS; crash/destroyed 12-17-59
56-1474 498th FIS; 37th FIS; 331st FIS; 460th FIS; OR ANG; CA ANG; PA ANG; drone
56-1475 37th FIS; 438th FIS; 59th FIS; 4780th ADW; SAALC; TX ANG; PA ANG; drone
56-1476 438th FIS; 59th FIS; MN ANG; PA ANG; OR ANG; drone
56-1477 438th FIS; 323rd FIS; 59th FIS; crash/destroyed 12-13-62
56-1478 11th FIS; 331st FIS; 460th FIS; OR ANG; drone
56-1479 37th FIS; 498th FIS; crash/destroyed 11-3-59
56-1480 37th FIS; 438th FIS; 59th FIS; ND ANG; PA ANG; drone
56-1481 11th FIS; 331st FIS; TX ANG; crash/destroyed 8-21-68
56-1482 11th FIS; ND ANG; TX ANG; NMIMT
56-1483 438th FIS; 87th FIS; 59th FIS; MN ANG; 509th FIS; MASDC
56-1484 438th FIS; crash/destroyed 5-14-59
56-1485 37th FIS; 11th FIS; 332nd FIS; PA ANG; MASDC Nov. 1969
56-1486 11th FIS; PA ANG; CA ANG; MASDC April 1970
56-1487 11th FIS; 332nd FIS; 57th FIS; MT ANG; LA ANG; TX ANG; drone
56-1488 438th FIS; 59th FIS; MN ANG; drone
56-1489 11th FIS; 59th FIS; MN ANG; ID ANG; CT ANG; VT ANG; WI ANG; drone
56-1490 11th FIS; 59th FIS; 327th FG; 4780th ADW; drone
56-1491 438th FIS; 59th FIS; 327th FG; TX ANG; 4780th ADW; 82nd FIS; scrapped at Naha AB 1971
56-1492 11th FIS; 59th FIS; 4756th ADG/4750th TG/ADWC; shot down by F-106 mid 1965
56-1493 11th FIS; 59th FIS; 327th FG; 4780th ADW; 509th FIS; 82nd FIS; scrapped at Naha AB 1971
56-1494 11th FIS; 59th FIS; 332nd FIS; MN ANG; PA ANG; drone
56-1495 498th FIS; 37th FIS; 331st FIS; 460th FIS; OR ANG; drone
56-1496 498th FIS; 37th FIS; 331st FIS; 460th FIS; OR ANG; CA ANG; drone
56-1497 11th FIS; 59th FIS; 325th FIS; 326th FIS; 4780th ADW; SC ANG; drone
56-1498 438th FIS; crash/destroyed 11-10-58
56-1499 438th FIS; 59th FIS; 509th FIS; 4780th ADW; CA ANG
56-1500 11th FIS; 64th FIS; crash/destroyed 4-26-62
56-1501 327th FIS; PA ANG; MASDC Nov. 1969
56-1502 11th FIS; 331st FIS; 57th FIS; MN ANG; TX ANG; SD ANG; ND ANG; Fargo Airport, ND display
56-1503 11th FIS; CA ANG; TX ANG; MASDC July 1970
56-1504 498th FIS; 37th FIS; 331st FIS
56-1505 438th FIS; 59th FIS; 57th FIS; MN ANG; Minot AFB display
56-1506 59th FIS; 325th FIS; 4780th ADW; 82nd FIS; scrapped at Naha AB 1971
56-1507 498th FIS; 37th FIS; 331st FIS; 460th FIS; 82nd FIS; scrapped Naha AB 1971
56-1508 11th FIS; TX ANG; crash/destroyed 2-23-63
56-1509 11th FIS; 64th FIS; 326th FIS; 4780th ADW; PA ANG; drone
56-1510 11th FIS
56-1511 498th FIS; 37th FIS; crash/destroyed 9-29-59
56-1512 37th FIS; 4756th ADW; 438th FIS; 323rd FIS; 59th FIS; 4780th ADW; 525th FIS; MT ANG; NY ANG; CA ANG; drone
56-1513 498th FIS; 37th FIS; 331st FIS
56-1514 4756th ADW; 11th FIS; 331st FIS; TX ANG; MASDC Jan. 1970
56-1515 11th FIS; 64th FIS; 326th FIS; 4780th ADW; PA ANG; MASDC 4-9-98
56-1516 325th FIS; 4756th ADW; ADWC; 498th FIS; 37th FIS; 331st FIS; 4780th ADW; 64th FIS

56-1517 11th FIS; 64th FIS; 326th FIS; 4780th ADW; ADWC; SC ANG; crash/destroyed 12-5-72
56-1518 4756th ADW; 18th FIS; LA ANG; ND ANG; TX ANG; MASDC Nov. 1969
57-0770 82nd FIS; 498th FIS; 37th FIS; 331st FIS; 482nd FIS; VT ANG; NY ANG; drone
57-0771 498th FIS; 37th FIS; 331st FIS; 482nd FIS; VT ANG; SC ANG; drone
57-0772 498th FIS; 37th FIS; 331st FIS; 460th FIS; 82nd FIS; scrapped at Naha AB 1971
57-0773 498th FIS; 37th FIS; 331st FIS; 460th FIS; crash/destroyed 10-22-64
57-0774 498th FIS; 37th FIS; 331st FIS; 460th FIS; 82nd FIS; scrapped at Naha AB 1971
57-0775 498th FIS; 37th FIS; 460th FIS; TN ANG; SC ANG; CA ANG; MASDC 1971; Clovis Park, CA display
57-0776 498th FIS; 37th FIS; 331st FIS; 460th FIS; 460th FIS; OR ANG; CA ANG; drone
57-0777 498th FIS; 37th FIS; 331st FIS; China Lake NWC; 460th FIS; OR ANG; NAS China Lake, CA display
57-0778 498th FIS; 37th FIS; 331st FIS; 460th FIS; 82nd FIS; scrapped at Naha AB 1971
57-0779 498th FIS; 37th FIS; 331st FIS; 460th FIS; 82nd FIS; scrapped at Naha AB 1971
57-0780 498th FIS; 37th FIS; 331st FIS; 460th FIS; 82nd FIS; scrapped at Naha AB 1971
57-0781 498th FIS; 37th FIS; 331st FIS; 460th FIS; OR ANG; ID ANG; PA ANG; CA ANG; drone
57-0782 498th FIS; crash/destroyed 3-27-59
57-0783 498th FIS; 37th FIS; 331st FIS; 460th FIS; 82nd FIS; scrapped at Naha AB 1971
57-0784 498th FIS; 37th FIS; 331st FIS; 4780th ADW; 325th FIS; 82nd FIS; scrapped at Naha AB 1971
57-0785 498th FIS; 460th FIS; SC ANG; FL ANG; MASDC Nov. 1969
57-0786 325th FIS; 327th CAMS; 59th FIS; ND ANG; PA ANG; drone
57-0787 325th FIS; 327th CAMS; 326th FIS; 4780th ADW; PA ANG; drone
57-0788 71st FIS; 325th FIS; 327th CAMS; WA ANG; NY ANG; Westhampton Beach ANGB, NY display
57-0789 325th FIS; 327th CAMS; crash/destroyed 10-7-58
57-0790 325th FIS; 327th CAMS; crash/destroyed 1-7-62
57-0791 325th FIS; 327th CAMS; crash/destroyed 9-9-64
57-0792 325th FIS; 327th CAMS; 59th FIS; ND ANG; PA ANG; drone
57-0793 82nd FIS; 4750th TS; w/o 3-22-60
57-0794 325th FIS; 327th CAMS; 82nd FIS; scrapped at Naha AB 1971
57-0795 325th FIS; 327th CAMS; crash/destroyed 6-16-65
57-0796 325th FIS; 327th CAMS; 82nd FIS; scrapped at Naha AB 1971
57-0797 325th FIS; 327th CAMS; crash/destroyed 10-19-62
57-0798 325th FIS; 327th CAMS; crash/destroyed 10-16-64
57-0799 325th FIS; 327th CAMS; 82nd FIS; scrapped at Naha AB 1971
57-0800 325th FIS; 327th CAMS; 59th FIS; 317th FIS; 4780th ADW; CA ANG; 4756th ADW; OR ANG; PA ANG; drone
57-0801 325th FIS; 327th CAMS; 326th FIS; 4780th ADW; crash/destroyed 8-18
57-0802 329th FIS; 16th FIS; WA ANG; ID ANG; 82nd FIS; scrapped at Naha AB
57-0803 329th FIS; 460th FIS; 4780th ADW; SC ANG
57-0804 325th FIS; 327th CAMS; 82nd FIS; scrapped at Naha AB 1971
57-0805 325th FIS; 327th CAMS; crash/destroyed 8-4-58
57-0806 329th FIS; 82nd FIS; 460th FIS; OR ANG; CA ANG; drone
57-0807 82nd FIS; 325th FIS; 327th CAMS; 326th FIS; 4780th ADW; ID ANG; drone
57-0808 325th FIS; 327th CAMS; 326th FIS; 4780th ADW; drone
57-0809 325th FIS; 327th CAMS; crash/destroyed 12-16-62
57-0810 329th FIS; 460th FIS; 4780th ADW; SC ANG
57-0811 329th FIS; 460th FIS; TN ANG; PA ANG; CA ANG
57-0812 329th FIS; 82nd FIS; 325th FIS; 326th FIS; 4780th ADW; Kirtland AFB display
57-0813 329th FIS; 326th FIS; 4756th ADW; ADWC; PA ANG; drone
57-0814 325th FTS; 327th CAMS; 59th FIS; VT ANG; crash/destroyed 11-10-67
57-0815 325th FIS; 327th CAMS; 82nd FIS; scrapped at Naha AB 1971
57-0816 329th FIS; 482nd FIS; crash w/o 11-20-63
57-0817 329th FIS; 482nd' FIS; VT ANG; NY ANG; MT ANG
57-0818 329th FIS; 482nd FIS; 4756th ADW; ADWC; SC ANG
57-0819 329th FIS; damaged w/o 4-12-61
57-0820 329th FIS; 82nd FIS; crash/destroyed mid-air 9-13-63
57-0821 325th FIS; 327th CAMS; 326th FIS; 4780th ADW; WI ANG; MASDC 5-13-71
57-0822 329th FIS; 82nd FIS; 325th FIS; WI ANG; VT ANG; drone
57-0823 329th FIS; 482nd FIS; SC ANG; WA ANG; PA ANG; drone
57-0824 329th FIS; 482nd FIS; VT ANG; drone
57-0825 329th FIS; 482nd FIS; VT ANG; PA ANG; drone
57-0826 329th FIS; crash/destroyed 7-31-58
57-0827 329th FIS; 325th FIS; crash/destroyed 4-27-62
57-0828 329th FIS; 482nd FIS; VT ANG; ID ANG; WI ANG; SC ANG; drone
57-0829 325th FIS; 329th FIS; 327th CAMS; 326th FIS; 4780th ADW; ID ANG; drone
57-0830 482nd FIS; VT ANG; PA ANG; drone
57-0831 4756th ADW; 460th FIS; 4780th ADW; SC ANG; CA ANG; drone
57-0832 482nd FIS; 431st FIS; SC ANG; PA ANG; WI ANG; drone
57-0833 329th FIS; 460th FIS; OR ANG; VT ANG; Hill AFB display
57-0834 482nd FIS; VT ANG; OR ANG; drone

57-0835 329th FIS; 4756th ADW; USAF/Dept. of Energy tests; TX ANG; FAA; MASDC 12-15-70
57-0836 317th FIS; 3245th ABW (AFSC) as JEF-102A for air defense radar system trials; 526th FIS for 412-L-AWCS trials; 4780th ADW; MASDC 4-23-70
57-0837 329th FIS; 325th FIS; crash/destroyed 5-16-63
57-0838 329th FIS; 460th FIS; SC ANG; CA ANG; MASDC May 1971
57-0839 6520th TG; SC ANG; CA ANG; TX ANG; drone
57-0840 482nd FIS; 76th FIS; 82nd FIS; scrapped at Naha AB 1971
57-0841 482nd FIS; VT ANG; TX ANG; drone
57-0842 6520th TG as JF-102A; PA ANG; CA ANG
57-0843
57-0844 329th FIS; 326th FIS; crash/destroyed 5-13-63
57-0845 3245th ABW (AFSC) as JEF-102A; 526th FIS for 412-L-AWCS trials; Convair
57-0846 329th FIS; 460th FIS; 6520th TG as JF-102A; 3245th ABW (AFSC) as JF-102A; 526th FTS; SC ANG; CA ANG; Convair
57-0847 482nd FIS; SC ANG; PA ANG; WI ANG; drone
57-0848 325th FIS; 327th CAMS; 82nd FIS; scrapped at Naha AB 1971
57-0849 482nd FIS; 431st FIS; VT ANG; NY ANG; drone
57-0850 482nd FIS; crash/destroyed 10-10-62 57-
0851 4780th ADW; 325th FIS; 82nd FIS; scrapped at Naha AB 1971
57-0852 482nd FIS; VT ANG; drone
57-0853 482nd FIS; crash/destroyed 11-20-62
57-0854 482nd FIS; 76th FIS; SC ANG; VT ANG; ID ANG; drone
57-0855 482nd FIS; VT ANG; NY ANG; drone
57-0856 482nd FIS; VT ANG; CA ANG; drone
57-0857 482nd FIS; crash/destroyed 6-24-59
57-0858 482nd FIS; 4756th ADW; ADWC; Panama City, FL display
57-0859 482nd FIS; 460th FIS; SC ANG; MASDC Nov. 1969
57-0860 482nd FIS; VT ANG; drone
57-0861 4756th ADW; crash/destroyed Aug. 1958
57-0862 482nd FIS; crash/destroyed 7-7-65
57-0863 482nd FIS; VT ANG
57-0864 456th FIS; 82nd FIS; 4780th ADW; 64th FIS; 509th FIS
57-0865 456th FIS; 82nd FIS; scrapped at Naha AB 1971
57-0866 456th FIS; 82nd FIS; 460th FIS; OR ANG; MT ANG; crash/destroyed 6-24-71
57-0867 327th FIS; 4756th ADW; 460th FIS; TN ANG; PA ANG; MASDC Nov. 1969
57-0868 456th FIS; 82nd FIS; 326th FIS; 4756th ADW; ADWC; PA ANG; drone
57-0869 482nd FIS; 460th FIS; VT ANG; NY ANG; MT ANG; drone
57-0870 482nd FIS; ARDC; 57th FIS; SC ANG; CA ANG; drone
57-0871 482nd FIS; SC ANG; VT ANG; drone
57-0872 456th FIS; 82nd FIS; crash/destroyed 6-15-62
57-0873 482nd FIS; 4683rd ADW; crash/destroyed 12-14-62
57-0874 482nd FIS; crash/destroyed 2-6-65
57-0875 460th FIS; 456th FIS; crash/destroyed 12-22-59
57-0876 456th FIS; 460th FIS; TN ANG; PA ANG
57-0877 456th FIS; 82nd FIS; crash/destroyed 11-19-62
57-0878 456th FIS; 82nd FIS; 325th FIS; 326th FIS; 4780th ADW; drone
57-0879 456th FIS; 82nd FIS; 325th FIS; ND ANG; PA ANG; drone
57-0880 456th FIS; 82nd FIS; 460th FIS; 326th FIS; 64th FIS
57-0881 4756th ADW; 460th FIS; TN ANG; PA ANG; TX ANG; MASDC Nov. 1969
57-0882 456th FIS; 82nd FIS; scrapped at Naha AB 1971
57-0883 482nd FIS; SC ANG; VT ANG; OR ANG; WI ANG; drone
57-0884 456th FIS; 82nd FIS; scrapped at Naha AB 1971
57-0885 460th FIS; SC ANG
57-0886 456th FIS; 82nd FIS; crash/destroyed mid-air 4-22-65
57-0887 456th FIS; 82nd FIS; scrapped at Naha AB 1971
57-0888 456th FIS; 76th FIS; 82nd FIS; scrapped at Naha AB 1971
57-0889 456th FIS; 82nd FIS; 325th FIS; 59th FIS
57-0890 456th FIS; 82nd FIS
57-0891 456th FIS; 82nd FIS; scrapped at Naha AB 1971
57-0892 456th FIS; 82nd FIS; 460th FIS; OR ANG; WI ANG; PA ANG; crash/destroyed 1-24-74
57-0893 456th FIS; AFSC; 82nd FIS; accident/destroyed 4-21-67
57-0894 456th FIS; 82nd FIS; 460th FIS; OR ANG; CA ANG; drone
57-0895 456th FIS; 82nd FIS; scrapped at Naha AB 1971
57-0896 460th FIS; crash/destroyed 2-7-63
57-0897 460th FIS; crash/destroyed 1-28-60
57-0898 460th FIS; NASA; TN ANG; SC ANG; CA ANG; drone
57-0899 460th FIS; SC ANG; CA ANG; PA ANG; drone
57-0900 460th FIS; TN ANG; SC ANG; CA ANG; MASDC Nov. 1969
57-0901 460th FIS; crash/destroyed 12-15-60
57-0902 460th FIS; crash/destroyed 12-4-60
57-0903 460th FIS; SC ANG; CA ANG; drone
57-0904 460th FIS; TN ANG; TX ANG; crash/destroyed 1-28-65
57-0905 460th FIS; SC ANG; CA ANG; Ontario Airport, CA display
57-0906 460th FIS; 325th FIS; 59th FIS; 4756th ADW; ADWC; Robins AFB display
57-0907 460th FIS; 4780th ADW; SC ANG; CA ANG; drone
57-0908 460th FIS; SC ANG; CA ANG; drone
57-0909 460th FIS; 325th FIS; 326th FIS; 4780th ADW; SC ANG; VT ANG; drone

TF-102A

54-1351 ATC; 4780th ADW; OCAMA; became GTF-102A; Chanute AFB display; Selfridge museum

54-1352 4756th ADW; 4780th ADW; Turkey
54-1353 AFFTC as chase/test aircraft; Fox Field, Lancaster, CA
54-1354 AFFTC; AFSC; became NTF-102A; TX ANG; NMIMT
54-1355 73rd AD; crash/destroyed 1-18-58
54-1356 31st FIS; 73rd AD; 4750th ADG; 4780th ADW; NASA; became NTF-102A; MASDC 11-12-69
54-1357 317th FIS; 73rd AD
54-1358 323rd FIS; 86th FIS; 4756th ADW; 4780th ADW; NASA; became NTF-102A; MASDC 11-19-69
54-1359 327th FIS; 11th FIS; 31st FIS; 4756th ADG; 4780th ADW; became NTF-102A; MASDC 11-25-69
54-1360 327th FIS; 327th CAMS; 11th FIS; 323rd FIS; 325th FIS; 40th FIS; CT ANG; WI ANG; Turkey
54-1361 AFFTC; ADC; noise suppresser tests
54-1362 test
54-1363 318th FIS; 325th CAMS; 11th FIS; 32nd FIS; ME ANG; FL ANG; crash/destroyed 12-11-73
54-1364 2nd FIS; 526th FIS; 317th FIS; WA ANG; WI ANG; MASDC Dec. 1969
54-1365 2nd FIS; 52nd CAMS; 32nd FIS; ID ANG; MASDC Aug. 1974
54-1366 498th FIS; 526th FIS; 496th FIS; 525th FIS; ID ANG; CA ANG; Pima museum, AZ
54-1367 2nd FIS; 4780th ADW; 496th FIS; 431st FIS; 52nd CAMS; 526th FIS; CT ANG; CA ANG; MASDC Feb. 1975
54-1368 498th FIS; 329th FIS; ADWC; TX ANG; MASDC June 1974
54-1369 48th FIS; 4780th ADW; CA ANG; MASDC Dec. 1969
54-1370 438th FIS; 95th FIS; 11th FIS; 526th FIS; 32nd FIS; ME ANG; FL ANG; crashed into sea 2-10-71
55-4032 326th FIS; 68th FIS; automatic landing tests; CA ANG; Turkey
55-4033 48th FIS; ,Z ANG; Turkey
55-4034 438th FIS; 95th FIS; 497th FIS; 4780th ADW; ND ANG; SC ANG; MASDC March 1975
55-4035 326th FIS; 86th FIS; 87th FIS; 4780th ADW; SD ANG; CA ANG; Greece
55-4036 482nd FIS; 325th FIS; 327th CAMS; CA ANG; 509th FIS; accident w/o Thailand 1-15-67
55-4037 27th FIS; AZ ANG; ND ANG; MT ANG
55-4038 482nd FIS; 497th FIS; 431st FIS; crashed into sea 9-8-61
55-4039 27th FIS
55-4040 11th FIS; 4780th ADW; 509th FIS; MASDC 5-18-71
55-4041 82nd FIS; 497th FIS; 4780th ADW; HI ANG
55-4042 482nd FIS; 4780th ADW; HI ANG
55-4043 48th FIS; 82nd FIS; crash/destroyed 8-29-58
55-4044 332nd FIS; 438th FIS; 4780th ADW; MASDC 5-18-71
55-4045 64th FIS; 318th FIS; 325th CAMS; 431st FIS; 496th FIS; CA ANG; MASDC May 1974
55-4046 332nd FIS; 4756th ADW; 87th FIS; 82nd FIS; 4780th ADW; 40th FIS
55-4047 318th FIS; 325th CAMS
55-4048 5th FIS; 27th FIS; 4756th ADW; 87th FIS; 482nd FIS; SD ANG
55-4049 327th FIS; 64th FIS; 318th FIS; 325th CAMS; 460th FIS; TX ANG; PA ANG
55-4050 498th FIS; 325th FIS; 4780th ADW; MASDC 5-19-71
55-4051 86th FIS; 64th FIS; 4780th ADW; NMIMT
55-4052 86th FIS; 482nd FIS; 68th FIS; HI ANG; 4780th ADW; damaged by fire w/o 6-23-67
55-4053 497th FIS; 326th FIS; TX ANG; Turkey
55-4054 18th FIS; 4756th ADW; 4780th ADW; 332nd FIS; crash/destroyed 8-14-64
55-4055 18th FIS; 4756th ADW; 332nd FIS; 4780th ADW; MASDC 11-10-69
55-4056 61st FIS; 327th CAMS; 47th FIS; 4780th ADW; crash/destroyed 4-30-65
55-4057 332nd FIS; 4756th ADW; 47th FIS; TX ANG; MASDC Oct. 1971
55-4058 64th FIS; 325th CAMS; 325th FIS; 4780th ADW; MASDC 11-14-69
55-4059 4756th ADW; 18th FIS; 496th FIS; 526th FIS; crash/destroyed 6-8-67
56-2317 61st FIS; 327th CAMS; 37th FIS; 59th FIS; 4780th ADW; 4756th ADW; ADWC; 317th FIS; 331st FIS; TX ANG; ID ANG; Tyndall AFB display; Belleville, MI museum
56-2318 4756th ADW; 18th FIS; 76th FIS; crash/destroyed 7-26-62
56-2319 4756th ADW; 27th FIS; 48th FIS; 4780th ADW
56-2320 438th FIS; 4780th ADW; MASDC 11-17-69
56-2321 4756th ADW; 71st FIS; 4780th ADW; MASDC 11-20-69
56-2322 4756th ADW; 456th FIS; 326th FIS; 4780th ADW; MASDC 6-17-70
56-2323 71st FIS; 76th FIS; 4780th ADW; 4756th ADW; ADWC; TX ANG; MASDC Aug. 1974
56-2324 4756th ADW; 4th FIS; CT ANG; NMIMT
56-2325 37th FIS; 40th FIS; LA ANG; AZ ANG; Turkey
56-2326 61st FIS; 327th CAMS; 68th FIS; CT ANG; Greece
56-2327 37th FIS; 331st FIS; CA ANG; Greece
56-2328 332nd FIS; 16th FIS; 509th FIS
56-2329 18th FIS; 332nd FIS; 496th FIS; 431st FIS; 526th FIS; 32nd FIS; CT ANG; FL ANG; MASDC Sept. 1974
56-2330 317th FIS; WI ANG; MASDC Oct. 1974
56-2331 317th FIS; 4780th ADW; 525th FIS; 4781st CCTS; SC ANG; FL ANG; MASDC Feb. 1974
56-2332 317th FIS; WI ANG; MASDC Oct. 1974
56-2333 317th FIS; 4780th ADW; ADWC; 525th FIS; 496th FIS; MT ANG; TX ANG; MASDC Aug. 1974; Eglin AFB display
56-2334 323rd FIS; LA ANG; CT ANG; Turkey

56-2335 332nd FIS; 16th FIS; 509th FIS; Greece
56-2336 323rd FIS; FL ANG; TX ANG; MASDC Aug, 1971
56-2337 327th FIS; 482nd FIS; 4780th ADW; TX ANG; Lackland AFB display
56-2338 327th FIS; 332nd FIS; 4780th ADW; crash/destroyed 7-3-63
56-2339 27th FIS; 87th FIS; 526th FIS; MT ANG; SC ANG; TX ANG; MASDC 1972
56-2340 18th FIS; TX ANG; crash/destroyed
56-2341 18th FIS; 87th FIS; 4780th ADW; crash/destroyed 1-12-67
56-2342 332nd FIS; 18th FIS; 325th FIS; 4780th ADW; OR ANG; Turkey
56-2343 27th FIS; 325th FIS; 4780th ADW
56-2344 37th FIS; 331st FIS; 4756th ADG; ADWC; VT ANG; SC ANG; MASDC 1975
56-2345 438th FIS; 71st FIS; 82nd FIS; 325th FIS; 4756th ADW; ADWC; 4750th TS
56-2346 327th FIS; 332nd FIS; 329th FIS; 4683rd ADW; 64th FIS; WA ANG; PA ANG; Ft. Indiantown Gap, PA display
56-2347 329th FIS; 331st FIS; 4756th ADW; ADWC; TX ANG
56-2348 47th FIS; 4756th ADW; FL ANG; CA ANG; MASDC July 1971
56-2349 2nd FIS; 52nd CAMS; 5th FIS; 326th FIS; 4756th ADW; TX ANG; MASDC June 1974
56-2350 47th FIS; 482nd FIS; 4756th ADW; 59th FIS; ID ANG; MASDC 10-29-75
56-2351 2nd FIS; 52nd CAMS; 48th FIS; 4780th ADW; 4756th ADW; MASDC 5-19-71
56-2352 2nd FIS; 52nd CAMS; 332nd FIS; 4683rd ADW; 460th FIS ID ANG; OR ANG; CA ANG; MASDC Oct. 1965; Kelly AFB, TX display
56-2353 456th FIS; 82nd FIS; 326th FIS; 59th FIS; 460th FIS; MN ANG; MT ANG; NY ANG; CA ANG; WI ANG; Volk Field, WI display
56-2354 47th FIS; 71st FIS; 1st CAMS; crash/destroyed 11-6-59
56-2355 327th FIS; 318th FIS; 460th FIS; TN ANG; VT ANG; PA ANG; Turkey
56-2356 456th FIS; 57th FIS; 4756th ADW; MASDC 5-1-73

56-2357 95th FIS; 57th FIS; crash/destroyed 12-16-58
56-2358 456th FIS; 82nd FIS; 325th CAMS; 318th FIS; 64th FIS; 482nd FIS; VT ANG; ME ANG; MASDC July 1974
56-2359 460th FIS; 4780th ADW; crash/destroyed 6-24-59
56-2360 318th FIS; crash/destroyed 6-13-58
56-2361 11th FIS; 59th FIS; crash/destroyed 8-1-61
56-2362 326th FIS; 4756th ADW; 325th FIS; 4780th ADW; 509th FIS; 82nd FIS; scrapped at Naha AB 1971
56-2363 498th FIS; 460th FIS; SC ANG; MASDC Aug. 1971
56-2364 48th FIS; 64th FIS; 59th FIS; MN ANG; MT ANG; NY ANG; MASDC May 1975; Sacramento Airport, CA display
56-2365 86th FIS; crash/destroyed 6-10-59
56-2366 82nd FIS; 4756th ADW; 64th FIS; scrapped at Naha AB 1971
56-2367 4756th ADW; 57th FIS; CA ANG; MASDC 5-3-73
56-2368 4756th ADW; 4780th ADW; 331st FIS; MN ANG; Turkey
56-2369 4756th ADW; 4780th ADW; MASDC 5-19-71
56-2370 325th FIS; 64th FIS; 4756th ADW; ADWC; TX ANG; MASDC Apr. 1975
56-2371 4756th ADW; 4780th ADW; crash/destroyed 5-14-64
56-2372 4756th ADW; ND ANG; CA ANG; MASDC Apr. 1972
56-2373 4756th ADW; 318th FIS; 4780th ADW; 325th FIS; 64th FIS; 82nd FIS; scrapped at Naha AB 1971
56-2374 4756th ADW; 59th FIS; 4780th ADW; MASDC 5-19-74
56-2375 4756th ADW; 3555th FTW (ATC); 4780th ADW; TX ANG; MASDC May 1971
56-2376 4756th ADW; 482nd FIS; ADWC; TX ANG; MASDC Oct. 1075
56-2377 4756th ADW; 4780th ADW; crash/destroyed 3-22-65
56-2378 4756th ADW; 4780th ADW; VT ANG; MASDC 5-19-71
56-2379 4756th ADW; 4780th ADW; IWS; TX ANG; MASDC March 1971

Appendix D
PAVE DEUCE Aircraft

TYPE	DRONE NO.	USAF S/N	IN PROGRAM	SORTIES	REMARKS
QF-102A	601	56-1475	4-20-73	2	crashed Holloman 1-31-75
	602	56-1347	4-20-73		
	603	56-1400	4-22-73		
	501	56-1443	2-23-76		became QF-102B
	502	56-1081	5-20-75		became QF-102B
PQM-102A	604	57-0870	6-11-73	3	destroyed Holloman 12-17-74
	605	56-1419	6-11-73	9	destroyed Tyndall 7-18-75
	606	56-1418	7-6-73	7	destroyed Holloman 12-11-74
	607	56-1487	7-11-73	1	destroyed Holloman 1-31-75
	608	56-1401	8-14-73	1	crashed Holloman 8-20-75
	609	56-1330	9-13-74	5	destroyed Holloman 8-28-75
	610	56-1480	9-13-74	11	destroyed Holloman 5-11-75
	611	56-1489	9-16-74	1	operational loss Holloman 9-17-75
	612	57-0894	9-16-74	6	destroyed Holloman 11-23-76
	613	57-0883	9-20-74	3	destroyed Tyndall 2-26-76
	614	56-1512	9-30-74	10	destroyed Tyndall 5-25-76
	615	57-0771	10-1-74	4	destroyed Holloman 2-16-77
	616	56-1349	10-7-74	3	destroyed Holloman 10-21-77
	617	57-0855	10-25-74	1	destroyed Holloman 3-4-77
	618	56-1509	11-13-74	4	destroyed Tyndall 6-10-76
	619	56-1460	1-6-75	8	destroyed Holloman 1-12-78
	620	56-1466	12-31-75	8	destroyed 11-7-77
	621	56-1360	12-31-75	5	destroyed Holloman 3-8-77
	622	57-0800	1-27-75		
	623	57-0776	1-27-75	4	destroyed Holloman 4-30-82
	624	56-1323	1-21-75		
	625	56-1336	1-28-75	6	destroyed Holloman 3-18-77
	626	56-1325	1-28-75	1	destroyed Holloman 10-2-76
	627	56-1434	1-28-75		pre-delivery crash 1976
	628	56-1496	1-28-75	1	destroyed Holloman 8-21-76
	629	57-0856	1-29-75	1	destroyed Holloman 10-15-76
	630	56-1474	2-10-75	3	destroyed Holloman 2-26-77
	631	56-1329	2-10-75	5	destroyed Holloman 7-21-80
	632	56-1398	2-21-75		
	633	57-0869	3-31-75	1	destroyed Holloman 4-24-78
	634	57-0781	2-4-75		
	635	56-1384	4-15-75	3	destroyed Holloman 2-8-78
	636	57-0770	12-31-75	1	destroyed Holloman 11-3-77
	637	57-0849	5-1-75	4	destroyed Holloman 8-9-78
	638	57-0854	10-1-75	2	destroyed Holloman 9-17-78
	639	56-1380	5-13-75	3	self-destructed Holloman 1977
	640	57-0813	5-13-75	3	destroyed Holloman 8-29-78
	641	57-0830	5-15-75	1	destroyed Holloman 9-17-78
	642	56-1331	12-31-75	7	destroyed Tyndall 9-18-78
	643	57-0807	12-31-75	5	destroyed Holloman 10-24-78
	644	56-1426	12-31-75	16	destroyed Holloman 11-17-81
	645	57-0786	5-29-75	3	destroyed Tyndall 8-30-78
	646	56-1340	5-27-75	1	destroyed Tyndall g-20-78
	647	57-0792	6-3-75	5	destroyed Holloman 10-12-79
	648	57-0823	6-3-75	1	destroyed Tyndall 8-29-78
	649	56-1046	1-15-75	2	destroyed Tyndall 2-15-79
	650	54-1385	8-6-75	5	destroyed Holloman 11-9-78
	651	56-1057	8-6-75	1	destroyed Holloman 9-15-78
	652	54-1407	9-4-75	1	destroyed Holloman 9-25-79
	653	54-1393	12-31-75	1	destroyed Holloman 11-13-78
	654	56-1208	12-31-75	2	crashed Holloman 11-17-78
	655	56-1054	10-1-75	10	destroyed Tyndall 6-19-79
	656	56-1061	10-7-75	2	crashed Holloman 11-13-79
	657	54-1399	10-7-75	13	destroyed Tyndall 5-31-79
	658	55-3427	10-8-75	1	destroyed Tyndall 12-12-78
	659	56-1255	8-6-75	6	destroyed Tyndall 6-14-79
	660	57-0832	12-8-76		
	661	54-1406	10-20-75	2	destroyed Tyndall 9-4-77
	662	55-3447	10-21-75	9	destroyed Tyndall 6-7-79
	663	57-0829	12-31-75	8	destroyed Holloman 4-23-81
	664	57-0825	12-8-76	11	destroyed Tyndall 6-5-79
	665	57-0847	2-2-77	2	crashed Holloman 8-28-79
	666	56-1048	2-2-77	3	destroyed Tyndall 8-29-79
	667	56-1055	2-9-77	4	destroyed Holloman 1-24-80
	700	56-1256	5-20-77	1	destroyed Tyndall 5-5-81

PQM-102B	701	56-10-77	5-15-78	2	destroyed Tyndall 8-8-79
	702	56-1211	5-15-78	1	destroyed Tyndall 8-30-79
	703	56-1101	6-27-78	2	destroyed Tyndall 4-17-80
	704	56-1107	6-28-78	1	destroyed Tyndall 1-30-80
	705	56-1210	6-29-78	1	destroyed Tyndall 12-20-79
	706	56-1181	7-14-78	9	crashed Holloman 7-16-80
	707	56-1184	7-14-78		
	708	56-1196	8-16-78		
	709	56-1263	8-17-78	1	crashed Holloman 2-29-80
	710	56-1257	8-23-78	2	destroyed Tyndall 5-15-80
	711	56-1121	9-26-78	6	destroyed Tyndall 4-14-81
	712	56-1111	9-27-78	1	destroyed Tyndall 5-15-80
	713	56-1006	9-29-78	4	destroyed Tyndall 12-4-80
	714	56-1093	10-18-78	7	destroyed Tyndall 1-8-81
	715	56-1272	10-26-78	2	destroyed Holloman 8-3-81
	716	56-1274	10-27-78	5	operational loss Tyndall 8-9-79
	717	56-0993	11-2-78	3	destroyed Tyndall 1-29-81
	718	56-1193	11-20-78	6	destroyed Tyndall 5-29-81
	719	56-1277	11-27-78	4	destroyed Tyndall 8-14-80
	720	56-1213	11-30-78	3	destroyed Tyndall 5-28-81
	721	56-1086	12-21-78	1	destroyed Holloman 3-15-80
	722	56-1130	1-19-79	5	destroyed Tyndall 8-21-79
	723	56-1014	1-23-79	2	destroyed Tyndall 8-16-79
	724	56-1098	1-24-79	1	destroyed Tyndall 9-10-81
	725	56-1269	1-31-79	1	destroyed Tyndall 7-26-79
	726	56-1284	2-8-79	4	destroyed Tyndall 12-4-80
	727	56-1083	2-16-79	3	destroyed Tyndall 1-9-81
	728	56-1278	3-8-79	4	destroyed Tyndall 1-9-81
	729	56-1472	3-9-79	1	destroyed Tyndall 9-6-79
	730	56-1337	3-30-79	6	destroyed Tyndall 12-18-81
	731	56-1241	4-6-79	6	destroyed Tyndall 4-17-80
	732	56-1427	4-18-79	1	destroyed Tyndall 10-18-79
	733	56-1476	4-19-79	3	destroyed Tyndall 9-9-79
	734	57-0806	4-26-79	20	destroyed Tyndal]. 9-3-81
	735	56-1032	5-3-79	2	crashed Tyndall 11-14-79
	736	56-1453	5-11-79	1	destroyed Tyndall 1-16-80
	737	57-0841	5-21-79	1	destroyed Tyndall 1-23-80
	738	56-1346	5-23-79	1	destroyed Tyndall 9-17-81
	739	56-1376	6-5-79	9	destroyed Tyndall 1-15-82
	740	56-1294	6-12-79	2	destroyed Tyndall 6-5-80
	741	55-3449	6-13-79	2	destroyed Tyndall 7-1-81
	742	56-1267	6-14-79	1	destroyed Tyndall 7-26-79
	743	56-1163	6-26-79	4	destroyed Tyndall 2-18-82
	744	56-1270	7-10-79	2	destroyed Tyndall 4-3-80
	745	56-1254	7-20-79		
	746	56-0982	7-25-79	2	crashed Holloman 6-5-81
	747	56-0980	7-27-79	3	destroyed Tyndall 6-5-80
	748	56-1223	7-31-79		
	749	56-1205	8-1-79	13	destroyed Tyndall 4-9-81
	750	56-1023	8-8-79	3	destroyed Tyndall 8-19-80
	751	56-1248	7-21-79	2	destroyed Holloman 3-17-80
	752	56-1291	7-23-79	3	destroyed Tyndall 2-23-82
	753	56-1043	8-28-79	2	destroyed Holloman 7-23-80
	754	56-1013	9-11-79	1	destroyed Holloman 3-17-80
	755	56-1188	9-14-79	1	destroyed Holloman 4-18-80
	756	56-1290	9-21-79	1	operational loss Holloman 8-15-80
	757	56-1044	9-26-79		crashed Nellis 7-8-80
	758	56-1063	10-5-79	1	destroyed Holloman 7-18-80
	759	56-1028	10-15-79	2	destroyed Tyndall 6-3-80
	760	57-0828	10-22-79	1	destroyed Holloman 11-23-81
	761	56-1262	10-25-79		
	762	56-1206	10-27-79	4	destroyed Tyndall 2-11-82
	763	56-1430	11-26-79	1	destroyed Tyndall 2-18-82
	764	56-1244	11-28-79	1	destroyed Tyndall 2-11-82
	765	56-1227	11-29-79	2	destroyed Tyndall 5-21-80
	766	56-1127	11-30-7g	1	destroyed Tyndall 8-20-80
	767	56-1446	12-3-79	3	destroyed Tyndall 3-17-82
	768	56-0987	12-19-79	1	destroyed Holloman 8-18-80
	769	57-0879	12-20-79	2	destroyed Tyndall 6-18-80
	770	56-1457	1-2-80		
	771	56-1259	1-7-80	2	destroyed Holloman 7-20-82
	772	56-1366	1-25-80	1	crashed Tyndall 10-6-80
	773	56-1289	2-5-80		
	774	56-1356	2-12-80	1	operational loss Holloman 8-21-80
	775	56-1447	2-29-80	1	withdrawn from use 7-23-80
	776	56-1341	2-29-80	3	destroyed Holloman 8-1-80
	777	56-1319	3-5-80	1	destroyed Tyndall 2-3-81
	778	56-1295	3-14-80	3	destroyed Holloman 8-26-81
	779	56-1350	3-24-80	3	destroyed Holloman 9-30-81
	780	56-1050	3-28-80		
	781	56-1042	4-2-80	3	destroyed Tyndall 4-15-82
	782	57-0787	4-29-80	1	destroyed Tyndall 4-14-82
	783	56-1062	4-3-80		

784	57-0808	5-9-80	2	destroyed Tyndall 9-4-81
785	56-1367	5-27-80		
786	57-0899	5-29-80		
787	56-1292	5-30-80		
788	56-1096	6-7-80	3	destroyed Tyndall 1-12-82
789	56-1122	July 1980		
790	57-0909	7-19-80	2	destroyed Tyndall 3-3-82
791	56-1464	7-21-80	1	destroyed Tyndall 4-27-82
792	57-0871	7-22-80		
793	56-1175	July 1980		
794	56-1170	7-29-80	16	destroyed Tyndall 11-5-81
795	56-1253	8-12-80	4	destroyed Holloman 12-1-81
796	57-0839	8-15-80		
797	56-1082	9-5-80		
798	56-0983	9-26-80	2	destroyed Tyndall 6-10-82
799	56-1456	9-26-80		
800	56-1072	10-4-80		
801	57-0834	10-8-80		
802	56-1488	10-9-80		
803	56-1490	10-21-80	3	withdrawn from use Oct. 1982
804	56-1363	10-21-80		
805	56-1136	11-10-80	1	destroyed Tyndall 1-12-81
806	54-1395	11-20-80		
807	56-1449	11-22-80		
808	56-1343	11-27-80	5	destroyed Tyndall 5-6-82
809	56-1439	8-28-80	1	destroyed Tyndall 5-20-82
810	57-0878	Dec. 1980		
811	56-1249	12-17-80		
812	56-1373	1-6-81		
813	56-1135	1-20-81		
814	56-1471	1-27-81	1	withdrawn from use Oct. 1982
815	56-1351	1-28-81		
816	56-1345	1-29-81	1	destroyed Tyndall 6-29-82
817	57-0822	2-17-81		
818	57-0824	2-24-81		
819	57-0852	2-27-81		
820	56-1478	3-5-81	6	destroyed Tyndall 8-25-81
821	56-1236	3-20-81	1	destroyed Tyndall 9-2-81
822	56-1287	3-23-81	2	destroyed Tyndall 8-26-81
823	56-1314	4-1-81		
824	56-0978	4-3-81		
825	56-1250	4-15-81	8	destroyed Tyndall 1-28-82
826	56-1202	4-22-81		
827	56-1293	4-23-81		
828	56-0996	4-29-81		
829	56-1300	5-1-81	3	destroyed Tyndall 10-7-81
830	56-1215	5-12-81		crashed 10-15-81
831	56-1261	5-19-81		
832	56-1003	5-29-81	2	destroyed Tyndall 4-14-82
833	57-0860	6-5-81	4	destroyed Holloman 6-8-82
834	56-1495	6-11-81		
835	56-1497	6-12-81		
836	56-1069	6-22-81		
837	56-1391	June 1981		
838	57-0903	6-24-81	1	destroyed Tyndall 3-11-82
839	57-0898	6-25-81	2	destroyed Tyndall 4-13-82
840	56-1494	7-7-81		
841	57-0907	7-13-81		
842	57-0868	7-22-81	3	destroyed Tyndall 3-9-82
843	56-1433	7-23-81	1	destroyed Tyndall 4-2-82
844	57-0908	7-27-81	2	destroyed Tyndall 5-11-82
845	57-0831	7-29-81		

Bibliography

BOOKS

Anderton, David A. The History of the U.S. Air Force: Hamlyn-Aerospace, London, 1982

Andrade, John M. U.S. Military Aircraft Designations and Serials since 1909: Midland Counties Publications, 1979

Cloe, John Haile. Top Cover for America, The Air Force in Alaska 1920-1983: Pictorial Histories Publications Co., 1984

Dwiggins, Don. They Flew the Bendix Race

Francillon, Rene J. The United States Air National Guard: Aerospace Publishing, Ltd., 1993

Gooberman, Lawrence A. Operation Intercept: Schaffer Library

Gunston, Bill. Fighters of the Fifties: Patrick Stevens Limited, 1981

Hallion, Richard P. On the Frontier - Flight Research at Dryden, 1946-1981, 1984

Lake, Jon. McDonnell F-4 Phantom - Spirit in the Skies: Aerospace Publishing, Ltd., 1992

Lashmar, Paul. Spy Flights of the Cold War: Naval Institute Press, 1996

McGregor A. and D. Wellden. United States Air Force and Army Serial Batches since 1946: Mach III Plus, 1993

McLaren, David R. Republic F-84: Schiffer Publishing, Ltd., 1998

Ragay, J.D. F-102 Delta Dagger in Europe: Squadron/Signal Publications, Inc., 1991

Schmidt, Harry P. Test Pilot: Mach 2 Books, 1997

GOVERNMENT PUBLICATIONS

AFSC Newsreview, April 1975

AMARC Document D003AF761 May 1998

Egress Systems Update, Aerospace Safety, April 1972

F-102 Operational Training Report

Handbook of Aerospace Defense Organization 1946-1986: Office of History, Air Force Space Command, 1986

History of the Air Force Flight Test Center: ARDC Historical Branch, 1954

History of 2652nd TFW: USAF-PACAF, 1965

SMAMA Flight Safety Material Deficiency Task Group Minutes, 1958

Targets Informational Manual; U.S. Army Missile Command, 1991

USAF Seventh Worldwide Weapons Meet - Project William Tell II, 1959

USAF T.O. 1F-102A-1

USAF T.O. 1F-102(Y)-3

PERIODICALS

Air Combat
Aircraft Illustrated
Airman
Air Progress
Airscoop
Alamagordo Daily News
The Alamo Eagle
Aviation Week
Aviation Week & Space Technology
British Aviation Review
Combat Aircraft
CONAD Panorama
Daily News (Anchorage, AK)
The Defender
Flight
Flight International
Friends Journal
GE Newsletter
Historian (MN ANG)
IPMS/USA Quarterly
IPMS Canada
Jet Letter (ND ANG)
Naval Aviation News
The Pontiac Press
Replica In Scale
Royal Air Force Flying Review
Saturday Evening Post

OTHER SOURCES OF VARIOUS DOCUMENTS

Consolidated Vultee Aircraft Corporation
Honeywell Defense Avionics Systems
National Aeronautics and Space Administration
Pima Air and Space Museum
San Diego Air and Space Museum
United States Air Force Museum